Mathematics Teaching

Mathematics Teaching On Target is a guidebook for improving mathematics teaching, based on the Teaching for Robust Understanding (TRU) Framework and its five dimensions – The Mathematics, Cognitive Demand, Equitable Access, Agency, Ownership, and Identity, and Formative Assessment. You'll be guided to refine your classroom activities across the five TRU dimensions, and your students will become more knowledgeable and resourceful thinkers and problem solvers.

Each chapter in *Mathematics Teaching On Target* introduces a set of easy-to-use questions for the hands-on improvement of lesson activities, such as:

- Think of an activity you use with your students. Is it as mathematically rich as it might be?
- Does it stretch your students in the right ways, inviting "productive struggle"?
- Can all students engage with it, in ways that allow them to grow as mathematical thinkers?
- What evidence will student work provide, helping you revise the activity so that it works better both in the moment and next time?

You'll find examples at the elementary, middle, and secondary levels for each dimension that show how addressing these questions can enhance mathematics instruction.

Ideal for your individual classroom, learning community, or district-level and wider professional development efforts, this book will enable you to help more students engage with mathematics in increasingly powerful ways. Beyond individual lessons, this book will also accelerate teacher development by helping you focus and reflect on what really counts in your instruction.

Alan Schoenfeld is a Distinguished Professor of Education and Mathematics at the University of California, Berkeley, USA.

Heather Fink is a postdoctoral researcher at Portland State University, USA.

Alyssa Sayavedra is an Assistant Professor of Elementary Education at California State University, Monterey Bay, USA.

Anna Weltman is a curriculum writer and researcher at the XQ Institute in Oakland, CA, USA.

Sandra Zuñiga-Ruiz is an Assistant Professor in the Teacher Education Department at San José State University, USA.

Studies in Mathematical Thinking and Learning
Alan H. Schoenfeld, Series Editor

Recent Publications

Hulbert/Petit/Ebby/Cunningham/Laird.
A Focus on Multiplication and Division: Bringing Research to the Classroom

Ebby/Hulbert/Broadhead.
*A Focus on Addition and Subtraction: Bringing Mathematics Education
Research to the Classroom*

Clements/Sarama.
Learning and Teaching Early Math: The Learning Trajectories, Third Edition

Akihiko Takahashi.
*Teaching Mathematics Through Problem-Solving: A Pedagogical
Approach from Japan*

Horn & Garner.
*Teacher Learning of Ambitious and Equitable Mathematics Instruction:
A Sociocultural Approach*

Petit/Laird/Ebby/Marsden.
*A Focus on Fractions: Bringing Mathematics Education Research to the
Classroom, Third Edition*

Schoenfeld/Fink/Sayavedra/Weltman/Zuñiga-Ruiz.
*Mathematics Teaching On Target: A Guide to Teaching for Robust
Understanding at All Grade Levels*

Schoenfeld/Fink/Zuñiga-Ruiz/Huang/Wei/Chirinda.
*Helping Students Become Powerful Mathematics Thinkers:
Case Studies of Teaching for Robust Understanding*

For a full list of titles in this series, please visit: https://www.routledge.
com/Studies-in-Mathematical-Thinking-and-Learning-Series/book-series/
LEASMTLS

Mathematics Teaching On Target

A Guide to Teaching for Robust Understanding at All Grade Levels

Alan Schoenfeld, Heather Fink, Alyssa Sayavedra,
Anna Weltman, and Sandra Zuñiga-Ruiz

Routledge
Taylor & Francis Group

NEW YORK AND LONDON

Designed cover image: © Getty Images/Randy Faris

First published 2023
by Routledge
605 Third Avenue, New York, NY 10158

and by Routledge
4 Park Square, Milton Park, Abingdon, Oxon, OX14 4RN

Routledge is an imprint of the Taylor & Francis Group, an informa business

© 2023 Alan Schoenfeld, Heather Fink, Alyssa Sayavedra, Anna Weltman, and
Sandra Zuñiga-Ruiz

ISBN: 978-1-032-45419-1 (hbk)
ISBN: 978-1-032-44167-2 (pbk)
ISBN: 978-1-003-37690-3 (ebk)

DOI: 10.4324/9781003376903

Typeset in Palatino
by codeMantra

Contents

About the Authors

Alan Schoenfeld is a Distinguished Professor of Education and Mathematics at the University of California, Berkeley. His research focuses on mathematical thinking and problem solving, teaching, and learning. Throughout his career he has worked in partnership with school districts, teachers, instructional designers, and professional learning communities, with the goal of enriching mathematics instruction and supporting equitable and ambitious instruction. This book builds on more than four decades of such partnerships. (For more detail, see https://bse.berkeley.edu/alan-h-schoenfeld and https://truframework.org/).

Heather Fink is a postdoctoral researcher at Portland State University working on an NSF-funded project supporting equitable and ambitious teaching practices. Before earning her doctorate in mathematics education at University of California, Berkeley, she worked for 11 years as a middle school math teacher and instructional coach in the San Francisco Bay Area. Heather is a National Board Certified Teacher (early adolescence/mathematics) who is committed to promoting educational equity through her teaching and research by challenging persistent injustices and unfair power distributions in mathematics classrooms. Her research focuses on understanding how inequities are constructed through classroom interactions and how, at a micro-level, interactions shape the opportunities students have to engage in rich content and to build positive identities as thinkers, learners, and members of communities.

Alyssa Sayavedra is an Assistant Professor of Elementary Education at California State University, Monterey Bay. She earned her doctorate in science and mathematics education at the University of California, Berkeley in 2018. She taught high school mathematics in Oakland Unified School District for five years and worked as a teacher-researcher and youth development facilitator in Oakland for over a decade. Her research and teacher education work emphasizes culturally relevant mathematics teaching, mathematics classroom discourse, and building on the strengths of multilingual learners. For her dissertation research, she partnered with four experienced, Black mathematics teachers in Oakland to form a professional learning community focused on culturally relevant teaching that builds on students' mathematical ideas.

Anna Weltman is a curriculum writer and researcher at the XQ Institute in Oakland, CA. She has a doctorate in mathematics education from the University of California, Berkeley, and has worked as an elementary, middle, and high school math teacher. Anna's design and research work focuses on building math learning environments rich with meaningful challenge and opportunities for creativity, in which students truly feel that they belong. She is an award-winning author of several math activity books for children and is currently working on a project-based math curriculum for the high school math sequence.

Sandra Zuñiga-Ruiz is an Assistant Professor in the Teacher Education Department at San José State University. Prior to earning her doctorate in mathematics education at UC Berkeley, she taught college mathematics at universities and community colleges in the Central Coast of California with an emphasis on issues of equity and social justice. Her research and teacher education work takes an emancipatory approach that seeks to cultivate dignity affirming mathematical learning communities for teacher candidates and children in the K-12 educational system. Her dissertation work investigated how four prospective maestras Mexicanas committed to justice developed understandings related to race, justice, and mathematics through the co-construction of a counterspace via critical conversations, pláticas.

Preface

There are so many things to attend to when we teach. There's the content and the ideas underlying it. There is the diverse collection of students in the classroom, who come with varied strengths and knowledge. There are the opportunities the curriculum does and doesn't provide for students to connect to big ideas, and questions of how to organize classroom activities so that they provide meaningful learning opportunities for as many students as possible. There are interactions between students, the teacher, and the content, and the ways those interactions shape students' knowledge and identities. There's hearing what students say in the moment and reacting to it productively. That's a lot to manage in the service of ambitious and equitable instruction.

So where do we start? What can we do on a daily basis to make progress as teachers?

The first step is to know what counts in teaching, so we can plan and act wisely in the service of our students' learning. The next step is to have tools that we can use, with others or by ourselves, to plan for and reflect on our teaching in ways that help us learn, on a day by day basis.

That's what this book is about. It's based on the Teaching for Robust Understanding (TRU) framework, which identifies the five main dimensions of classroom practice. These five dimensions help us attend to rich content learning and to building and maintaining equitable learning environments. To the degree that things go well along the five dimensions of TRU, students emerge from the classroom being powerful and empowered thinkers and learners. The introduction to this book describes the five dimensions of mathematically productive classrooms.

But then what? How do those big ideas get translated into day-to-day practice? We offer specific tools, in the form of "targets," to support this process. Suppose, for example, you want to modify learning tasks so that they are enriched mathematically and provide more opportunities for more students to dig into them. Where do you start? We list some resources, and some ideas. Then we get to the targets, which are aligned with the five dimensions of the TRU Framework.

On the outer rings of the targets are descriptions of classroom attributes and activities that are commonly found in mathematics lessons, but that, with some adjustments, hold the potential to support more equitable and ambitious learning opportunities. For example, on the outer ring of one of the targets focused on mathematical content you'll find "Activity focuses on a single procedure or concept in isolation." As you move toward the center of that target, the attributes listed describe increasingly powerful opportunities for student learning. So, on the same target you'll find "Activity elicits a connection between a method and the underlying concept" and then "Activity supports the development of a "big idea" at grade level" closer to the center of the target. Similarly, an outer-ring description on the first target concerned with issues of student agency, ownership, and identity is "Mathematics is experienced as abstract and disconnected from personal experience or knowledge," and inner-ring descriptions include "Students have the opportunity to connect classroom activities with personal or cultural knowledge" and "Curricular materials build on student experiences."

Now, imagine yourself thinking about a task or activity you plan to use. Each target helps you think about the current attributes of that task or activity, and, possibly, directions in which it might be enriched.[1] So, the targets can help in planning. They also help in review. After you've taught the lesson, you (and perhaps colleagues or coaches, who either visit the class or watch videos of it) can debrief, reflecting on what seemed to work well and for whom, and what might be done differently next time.

Over time, and especially with the help of supportive colleagues, two things will happen. First, the targets will provide support for continuous improvement – you'll see your classroom practices and activities getting richer, and your students becoming more mathematically powerful and empowered. Second, the ideas behind the targets will soon become second nature. This will make it easier to plan for your teaching, because you'll be focusing more explicitly on what counts. This becomes a natural part of thinking about your teaching.

Note

1 The goal isn't to get to the center of the target with every task or activity, but to enrich tasks and activities when it's possible. No task or activity is perfect, and we often make trade-offs in instruction. The idea is to be mindful of possibilities, and to take advantage of them when we can.

1

Introduction

No two learning environments are the same – every classroom is shaped by the teacher's and students' personalities, history, dispositions and knowledge, and the ways they play out in interaction. Likewise, there is no one "right way" to teach. Two excellent teachers may, as matters of personal style, use very different approaches to orchestrate classroom interactions that provide safe spaces for rich, meaningful learning. Teaching is a human practice that should build authentically on students' and teachers' personal and cultural experiences.

At the same time, there *are* consistencies in ambitious and equitable instruction. The classrooms from which students emerge as knowledgeable, flexible, and resourceful thinkers and problem-solvers share some basic properties. These classrooms offer students truly rich content to engage with and forms of engagement that allow students to build deep understandings and to develop positive mathematical identities. They value the knowledge students bring into the classroom from their homes and communities and make classroom learning relevant to their real worlds. Specifically, they provide significant opportunities along the five dimensions of robust learning environments summarized in Figure 1.1.

Here's the core idea behind Teaching for Robust Understanding (TRU). If you consistently focus on key aspects of the five dimensions highlighted in Figure 1.1, your teaching becomes increasingly responsive to your students' thinking.

DOI: 10.4324/9781003376903-1

The Five Dimensions of Powerful Mathematics Classrooms

The Mathematics	Cognitive Demand	Equitable Access to Mathematics	Agency, Ownership, and Identity	Formative Assessment
The extent to which classroom activity structures provide opportunities for students to become knowledgeable, flexible, and resourceful mathematical thinkers. Discussions are focused and coherent, providing opportunities to learn mathematical ideas, techniques, and perspectives, make connections, and develop productive mathematical habits of mind.	*The extent to which students have opportunities to grapple with and make sense of important mathematical ideas and their use. Students learn best when they are challenged in ways that provide room and support for growth, with task difficulty ranging from moderate to demanding. The level of challenge should be conducive to what has been called "productive struggle."*	*The extent to which classroom activity structures invite and support the active engagement of all of the students in the classroom with the core mathematical content being addressed by the class. Classrooms in which a small number of students get most of the "air time" are not equitable, no matter how rich the content: all students need to be involved in meaningful ways.*	*The extent to which students are provided opportunities to "walk the walk and talk the talk" – to contribute to conversations about mathematical ideas, to build on others' ideas and have others build on theirs – in ways that contribute to their development of agency (the willingness to engage), their ownership over the content, and the development of positive identities as thinkers and learners.*	*The extent to which classroom activities elicit student thinking and subsequent interactions respond to those ideas, building on productive beginnings and addressing emerging misunderstandings. Powerful instruction "meets students where they are" and gives them opportunities to deepen their understandings.*

Figure 1.1 The five essential dimensions of mathematics classroom practice – the Teaching for Robust Understanding (TRU) mathematics framework.

There is plenty of research, and a lot of practical experience, to back up the ideas in Figure 1.1. If you're interested, see Schoenfeld (2013, 2014, 2015, 2018) for some of the research; these and related papers can be found at the TRU Framework website, https://truframework.org. For extensive detail on how the five dimensions play out in practice, see Schoenfeld et al. (2023). For nearly a decade, TRU has been used in professional development with partners around the US (see Schoenfeld et al., 2019) and internationally. We draw upon all of that research and practice here, but mostly as background. Our main goal is to offer practical tools to improve our teaching.

This document supports enhancing instruction along the five essential dimensions of learning environments. It provides a structure for planning and reflecting on classroom activities, whether we are doing it by ourselves, in pre-service teaching, in in-service teaching with a partner teacher or coach, or as part of a professional learning community (PLC). The idea is to deepen and enrich our instructional practices – to build on what we're doing well and to provide further opportunities for our students to develop into knowledgeable, resourceful, and flexible mathematical thinkers and problem-solvers.

At the most general level, key planning and reflection goals related to the five dimensions are:

– *Key Goal for Dimension 1: Enriching the Mathematics, both Content and Practices.* Can the tasks and activities planned for the lesson be made more mathematically deep and connected, providing increased opportunities for student understanding?
– *Key Goal for Dimension 2: Finding the Right Levels of Cognitive Demand.* Are there ways to open up the mathematics while maintaining mathematical richness, so that more students can build their understanding through "productive struggle?"
– *Key Goal for Dimension 3: Providing Meaningful Access to the Mathematics for all Students.* Can activities be modified in ways that build on students' strengths and knowledge, allowing more students to engage in mathematical sense making?
– *Key Goal for Dimension 4: Providing Opportunities for Students to See Themselves as Mathematically Powerful Thinkers and Problem-Solvers.* Can the mathematics be re-framed or the activities re-shaped, so that students have increased opportunities to reason, conjecture, and solve problems and to put forth ideas and refine their own and others' ideas, so they increasingly come to see themselves as mathematical sense makers?
– *Key Goal for Dimension 5: Making Student Thinking Public to Support Student Growth in Dimensions 1 Through 4.* Can we pose problems, or open up classroom discourse, in ways that bring more student thinking out into the open – and in doing so, provide increased opportunities to enrich each of the other dimensions?

On ambitious and equitable instruction

This book is aimed at supporting ambitious and equitable instruction. A key aspect of ambitious instruction is a consistent focus on rich, connected, and powerful mathematics. Then: who engages, in what ways? Our goal for equitable instruction is for each and every student to engage with rich, powerful, and connected mathematics in ways that are personally meaningful – in ways that connect to who they are, and in which they come to see themselves as mathematical sense makers. This has implications both for pedagogy and for the curriculum. We focus on opening up the mathematics we teach and the ways in which we teach it, so that all students are welcomed into mathematics and make it their own. For detail, see the chapters addressing Dimension 3 (Chapter 5) and Dimension 4 (Chapter 6).

Targets: our metaphor for improvement

We use the metaphor of targets to think about improving our teaching practice. Imagine a target with three rings. The outer ring lists practices or attributes that are fine on their own terms but hold potential for improvement. The middle ring lists practices or attributes that are richer than those in the outer ring – they might help students to make more mathematical connections or provide more opportunities for students to draw on what they know, explain their thoughts, etc. The inner ring describes the attributes of potentially richer tasks or activities.

Now, consider a task or activity you've been planning to use. Imagine identifying the attributes of the task/activity on the target. If most of the attributes that have been identified are on or near the outer ring of the target, then it's likely that there's room for improvement! How might the task be enhanced? The attributes listed on the inner rings provide ideas for you to consider. They may well suggest task modifications. When you identify the attributes of your revised task or activity, those marks may cluster more tightly around the inner circle.

Of course, teaching is multi-dimensional; we could hardly address all the issues we need to address with one target. This book offers 15 in all, 3 for each dimension of TRU. Each target addresses one specific dimension-aligned focal question. You'll find the three targets for each dimension toward the end of the chapter devoted to that dimension. (Or look toward the end of this introduction, where we provide a sample target for Dimension 5, Formative Assessment.)

Using the targets

Here's what we do in practice. We start with the current version of a learning task or activity. It may be taken from a textbook, borrowed from a colleague, or adapted from lessons we've used in the past. It might do pretty well on some TRU dimensions but maybe not as well on some others.

We start by looking at the math targets (Dimension 1 of TRU), one at a time. We take the activity and identify its current attributes on the first math target. Then we look at the inner rings of the target and look for suggestions about how the activity might be improved (consistent with time constraints and curricular goals). If opportunities exist, we think about specific task modifications. We do the same for math targets 2 and 3.

The results of this process should "land" closer to the center of the math targets. But that's just the starting point. Might there be ways to modify the task/activity or the way we introduce it, so that more students can engage productively with it? There might be a preliminary activity in which students read the task and explain it to each other or an approach that opens up the task for more reasoning and sense making. The idea is to adjust cognitive demand (TRU Dimension 2) so that the task is within reach. Practically, we proceed as we did with the math targets. We look to see where the activity lands on each of the three cognitive demand targets. We look to the inner rings of the targets for suggestions about ways to enhance productive struggle and then think about modifying the task accordingly.

At this point the question is: has the task or activity been crafted so that all students can engage with it meaningfully? The equitable access targets (TRU Dimension 3) help us review the task to see if there are "ways in" to the task for all students and if the way we've planned to use the activity provides equitable opportunities. Then we proceed as above.

Similarly, what opportunities does the task or activity provide for students to develop a sense of mathematical agency, to "own" the mathematics by having made sense of it themselves, and to build positive identities by way of being mathematical sense makers? (This is TRU Dimension 4, Agency, Ownership, and Identity, or AOI for short.) Does the task or activity relate to students' lives in potentially meaningful ways? Does it build on what they know? Does it provide opportunities for students to venture ideas safely and to engage in collaborative sense making? The three AOI targets help us see where the task or activity can be opened up for more meaningful engagement and sense making.[1]

Finally, there's the issue of bringing students' thinking out into the open and adjusting instruction to address the challenges or possibilities that have been revealed. (That's TRU Dimension 5, Formative Assessment.) Does the task or activity call for students to make conjectures, explain their thinking, or in other ways provide information about the state of their understanding? Does it make use of students as resources for each other? How might it be modified accordingly? We address these questions the same way we've addressed the others. The three formative assessment targets help us identify the attributes of the current (already modified) task or activity and consider possible modifications.

This sounds like a huge amount of work! But it doesn't have to be. The key is taking one step at a time, focusing on gradual improvement. Often people who are new to TRU start by focusing on just one or two dimensions. At first, there's a lot on each target and it takes time to work through the attributes. With a bit of practice, you will be able to scan a target and quickly identify places where the task or activity you're considering can be enhanced. Thinking this way soon becomes habitual, and you will find yourself making adjustments "in the moment" as well as in planning. You will also notice connections between the targets within and across dimensions, noting how making a change in task design or implementation can move you closer to the inner rings on multiple targets.

Using the TRU dimensions for planning and reflection: an example

In the following example we start with a typical textbook task and work to enrich it using ideas from the five TRU dimensions. We want to convey the spirit of the enterprise in this example, so we won't go into detail using the targets. (There are plenty of examples using the targets in what follows.) If you do look at the targets, though, you might find some of the task attributes that led to the task modifications we're suggesting.

In practice, you might or might not have the time to use a "fully enriched" version of this task – but that's not the point. Once you can envision the space of possibilities, you can decide which aspects of the expansion you can actually take on at a given time or in a given context.

Here's a task which you'll find in most secondary curricula.

T1:   Show that if you draw in the two diagonals of a rectangle, you divide the rectangle into four equal areas.

T1 is a good exercise, involving important mathematical ideas. But, can it be improved? For one thing, the result is given. That removes possibilities of exploration – of mystery, surprise, and discovery. A useful technique for opening up "show that" problems is to turn them into "is it true that?" problems. T2 offers a more open version of the task.

T2:   My friend Gloria says that if you draw in the two diagonals of a rectangle, you divide the rectangle into four equal areas. Is she right? How do you know?

In this case, Gloria is correct, so students wind up verifying something true that has been suggested to them. But what if Gloria is wrong or partially correct? For figures other than rectangles, she might be. See the examples below.

A wide-open version of the problem is,

T3:   Consider an arbitrary quadrilateral. Draw in the diagonals. Do they divide the quadrilateral into four equal parts? Why or why not?

This version may be *too* open! Think about the language in T3. Will students understand or be able to deal with the concept of an "arbitrary quadrilateral"? Some students might, but we want this problem to be accessible for all students (Dimension 3, Equitable Access). A lot depends on what the students know and how much scaffolding might be appropriate for them. Also, there's some very subtle mathematics involved in answering a question like T3. The full meaning of T3 is,

T3a: If you take any quadrilateral and draw in the two diagonals, the quadrilateral is divided into 4 parts. For what kinds of quadrilaterals must the four parts be of equal area? For what kinds of quadrilaterals is that not necessarily true (i.e., you can find examples where the pieces are not all of the same area)?

Teasing out the possibilities in this problem is quite a challenge! There are various figures to consider – parallelograms, rhombuses, trapezoids, and irregular figures. And, there are tacit rules. Are you going to claim that a statement is always, never, or sometimes true? The mathematician does different things in each case (providing a proof in the case of "always" or "never" statements, and giving examples of where it does and doesn't work in the "sometimes" case.) That's a mathematical can of worms, and you don't want to get into it unless you've spent a lot of time planning for it!

So, something like T4 might be more appropriate:

T4: My friend José says: "if you draw in the two diagonals of any quadrilateral, you divide the quadrilateral into four equal areas."
Is his statement always, sometimes, or never true? If you think his statement is always true or never true, then how would you convince someone else?
If you think his statement is sometimes true, could you identify all the cases of a quadrilateral where it is true/not true?

The mathematical content of the task has now been enriched. But how will students interact with the task, with you, and with one another? That's where dimensions 2 through 5 come in more fully.

Let's turn to cognitive demand and equitable access, Dimensions 2 and 3. Is it possible that T4 will have no entry points for some students? Might it be wise, for this particular class, to begin with T2 and then move to T4 as a generalization of it? Might it be wise to have some scaffolding questions in your back pocket, so that students know to look for examples of (say) "kites" where the four regions are unequal in area and kites where they are equal?

Then, let's think about student voice (Dimension 4, Agency, Ownership, and Identity). How might the task be set up so that every student has a chance to think things through, venture ideas, test them, and work through solid chains of reasoning? How might the students present and discuss what they find? Finally, consider formative assessment (Dimension 5). If you decide it's worth going with a task like T4, how can you set things up so that student thinking is brought out into the open, so that instruction can meet them where they are, enriching the activities along all of the dimensions?

Of course, if you pursue all five dimensions to their fullest, the task T1 evolves into a major lesson in and of itself[2]! At any given time, you might or might not want to take on part or all of this endeavor. It depends on your goals for students, time pressures, and what they seem ready for at the moment. But even if you decide you only want to expand a little bit, thinking about these issues helps to open up the space of mathematics your students get to experience.

Grappling with complexity

There's a *lot* to think about when you plan for or reflect on a lesson. The five TRU dimensions help to organize your reflections. Instead of wondering "just where do I start," you know that there are five large categories of issues to grapple with – the five dimensions of Figure 1.1, or, in terms of goals, the five key goals presented shortly after Figure 1.1.

It helps to know about all five dimensions because teaching involves all of them. In the example we just worked through, we began by focusing on enriching the task (Dimension 1, The Mathematics) but then grappled with issues of keeping the task within reach for all students (Dimensions 2 and 3, Cognitive Demand and Equitable Access) and providing more ways for students to engage in sense making and see themselves as capable learners and doers of mathematics (Dimension 4, AOI). Doing this effectively means having a good sense of what students are thinking (Dimension 5, Formative Assessment). That's true of all instruction.

To get enough of a sense of the TRU Framework to use it meaningfully, you might want to read through this book once before digging into any specific targets. You might also want to look at two tools built by the Teaching for Robust Understanding Project, the *TRU Math Conversation Guide* (Baldinger, E. Louie, N., and the Algebra Teaching Study and Mathematics Assessment Project, 2018) and the *TRU Math Observation Guide* (Schoenfeld, A. H., and the Teaching for Robust Understanding Project, 2018). The *Conversation Guide* provides brief descriptions of the five TRU dimensions and a set of planning and reflection questions. The *Observation Guide* provides some "look fors" that help to portray what we'd like to see happening in our classrooms. Both documents can be used to get an overview of "big picture" issues, and both can be used in an ongoing way for purposes of continual improvement. They are part of the TRU Math Suite, a collection of tools and research that can be found at the websites http://truframework.org and http://map.mathshell.org/. Finally, for a deep look into mathematics classrooms using TRU, there's *Helping Students Become Powerful Mathematics Thinkers: Case Studies of Teaching for Robust Understanding* (Schoenfeld et al., 2023).

Overall, this document has nine parts. Following this introduction there's a detailed example that shows how we use the targets to enhance instruction. Then there are five chapters that focus, in order, on the five TRU dimensions. We conclude with a list of useful classroom strategies and a set of references and resources.

Each of the chapters that focuses on a particular TRU dimension has three parts. The first part of the chapter introduces the dimension. It explains why it's important, identifies some useful techniques for making progress along that dimension, provides some detailed questions for deeper planning and review, and identifies some useful resources. It also breaks the big goal or question for the dimension into three subgoals or sub-questions – specific issues to focus on as you dig more deeply into the dimension. The second part of the chapter introduces the targets. A first figure highlights the three main sub-questions for that dimension. It's followed by three targets, each of which elaborates on one of the sub-questions. The third part provides examples at the elementary, middle, and high school levels of how we use the targets to enhance a typical curricular task.

To make things concrete, we will take a closer look at one of the targets associated with Dimension 5: Formative Assessment.

The big idea behind formative assessment is that what students know, when revealed, can serve as a base on which further understandings can be built. This is summarized as follows on the first formative assessment figure: "To what extent is students' mathematical thinking surfaced; to what extent does instruction build on student ideas when potentially valuable, or address emerging misunderstandings?" That top-level question is then broken down into three main sub-questions:

– In what ways are student ideas, strategies, and reasoning processes brought out into the open?
– In what ways are students' informal understandings and language use valued and built on?
– In what ways do whole class activities or group interactions support the refinement of student thinking?

The target in Figure 1.2 corresponds to the second of these three questions.

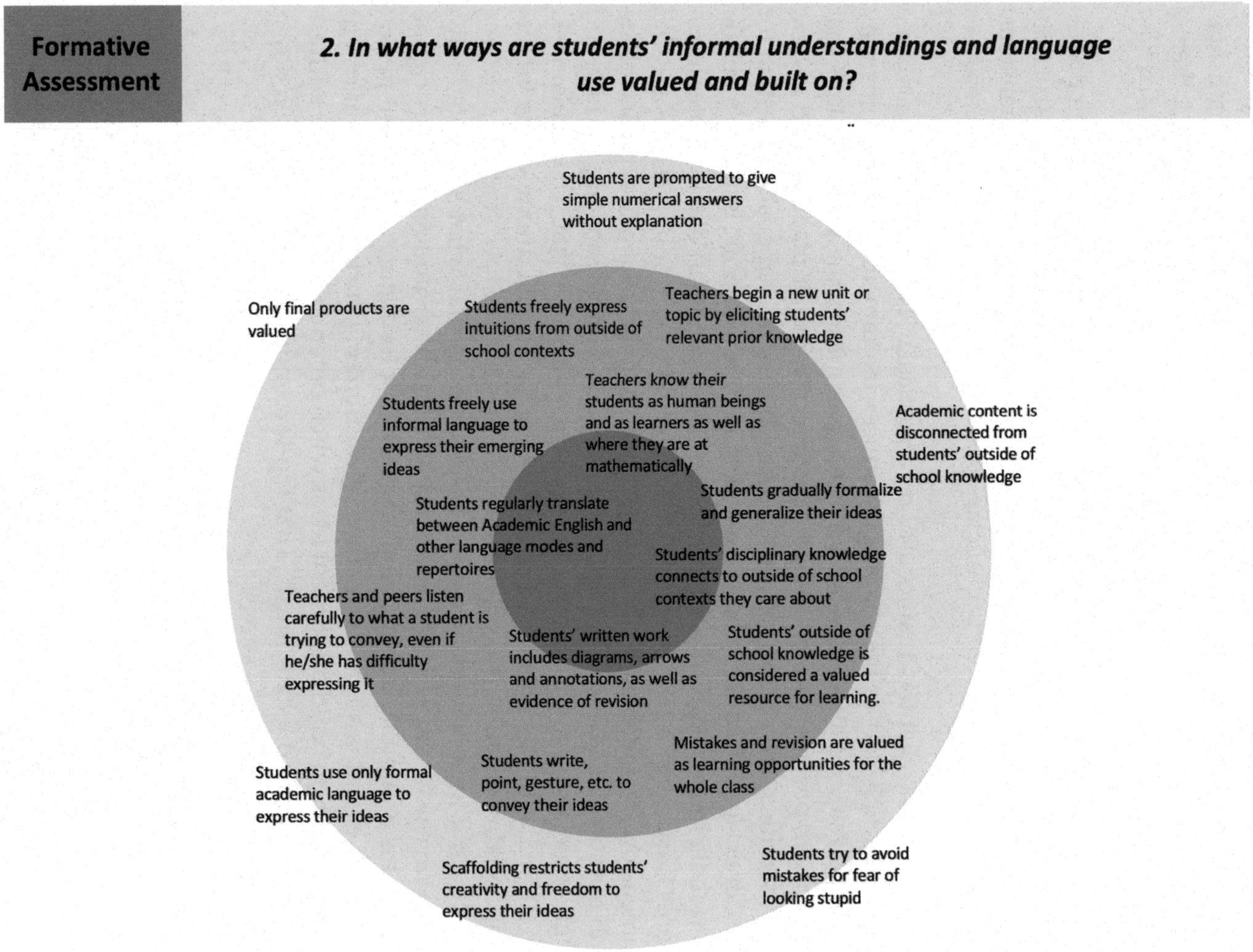

Formative Assessment

2. In what ways are students' informal understandings and language use valued and built on?

Students are prompted to give simple numerical answers without explanation

Teachers begin a new unit or topic by eliciting students' relevant prior knowledge

Academic content is disconnected from students' outside of school knowledge

Students freely express intuitions from outside of school contexts

Teachers know their students as human beings and as learners as well as where they are at mathematically

Students gradually formalize and generalize their ideas

Only final products are valued

Students freely use informal language to express their emerging ideas

Students' disciplinary knowledge connects to outside of school contexts they care about

Students' outside of school knowledge is considered a valued resource for learning.

Students regularly translate between Academic English and other language modes and repertoires

Students' written work includes diagrams, arrows and annotations, as well as evidence of revision

Mistakes and revision are valued as learning opportunities for the whole class

Students try to avoid mistakes for fear of looking stupid

Teachers and peers listen carefully to what a student is trying to convey, even if he/she has difficulty expressing it

Students write, point, gesture, etc. to convey their ideas

Students use only formal academic language to express their ideas

Scaffolding restricts students' creativity and freedom to express their ideas

Figure 1.2 The second target corresponding to formative assessment.

To work through just one example, let's look at the pathway that begins at the upper left-hand corner and moves toward the center of the target. On the outer ring we find, "Only formal academic language and final products are valued" – a situation that is problematic if we hope to build on what our students understand. Closer to the center we find, "Students freely use informal language to express their emerging ideas." Here we can imagine students thinking through their understandings, using the informal language at their disposal. This is certainly an enhancement, in that the students are now engaged in working through relevant ideas. But, we can go further: the goal is to have students become comfortable using formal mathematical/academic language to express *their* ideas. The third point along this trajectory, "Students regularly translate between Academic English and other language modes and repertoires," indicates that we are getting pretty close to that goal.

Now, suppose you're thinking about where the class is along this trajectory. Where is the class as a whole? Where are individual students? Can you think about how to move them, individually or collectively, closer to the center of the target? If that's a task you want to take on, the information in the section "What the formative assessment dimension involves" can help you do it. Make a habit of reflecting on the items on the formative assessment targets, and you'll get better at formative assessment. (And, at providing access and adjusting cognitive demand. In practice, the dimensions overlap.)

More broadly, the idea is that the comments on the targets represent typical attributes of curriculum tasks and classroom activities. If we can reflect on tasks and see where they "land" on the targets, then we can think about how we might alter them so that the revised versions land closer to the centers of the targets. Over time, this kind of reflection will result in richer raw materials to bring to the classroom.[3] In that way, this document is a tool for continuous improvement.

What to do first?

There's a lot to think about, even with just one target – and there are a whole bunch of targets! Whether you're working individually or (better) with a coach or partner, or as part of a PLC, taking on all of the targets at once may be too large a challenge. It may make sense to focus on a small number of targets at the beginning, until you feel comfortable with them and the progress you've made. You can add more targets as you become increasingly comfortable with their use.

When you're getting started, focusing on the math and one other dimension might be a good goal for a semester or perhaps even longer. Once thinking about those dimensions has become more or less habitual, you might add another dimension, while continuing to build on the progress you've made. Over time, as your repertoire grows and thinking in TRU terms becomes more natural, you'll have the ideas in your head. The targets always serve as tools to think with, however – we still use them in our practice.

Also, you'll see that we've included a blank target for each dimension. The idea is for you to customize that target and to focus on what matters to you. If there's something you really want to work on, think about stepping stones to what you'd like to see and write them in on the blank target. That way you can establish your own goals and work toward them. The TRU dimensions highlight "what counts," in broad terms. There's plenty of room for fine-tuning or for working on high priority goals within them.

Finally, please don't take the placement of particular items on the targets too literally! The closer to the center of a target an item is, the richer it is. Some items are long, so they spread over more than one ring of the target. Use your judgment to reflect on the items and on the tasks you're working to improve.

We hope that you will make the ideas in this book part of your repertoire and that using them will help to enrich your teaching, individually and in partnership with others.

Acknowledgments

The ideas for this book evolved over a number of years during conversations in the Teaching for Robust Understanding research group. Many students, visiting scholars, and R&D partners contributed to our thinking as this book took shape. Special thanks go to Anna Zarkh, who suggested the target metaphor.

This book was produced with the support of National Science Foundation grant 1503454, TRUmath and Lesson Study: "Supporting Fundamental and Sustainable Improvement in High School Mathematics Teaching" and the Bill and Melinda Gates Foundation grant OPP1115160, "Networked Improvement Community to Support Common Core State Standards Implementation." The authors and not the Foundations are responsible for the contents of the document.

Notes

1   It's important to note that opening up a task should not be taken to mean "simplifying." Tasks that provide opportunities for multiple solutions can both be more accessible (different students may find different ways to think about and approach the task) and richer (comparing and contrasting different approaches can illuminate important mathematical ideas).

2   We have such a lesson in mind. Take a look at Mathematics Assessment Project's Formative Assessment Lesson *Evaluating statements about length and area*, at http://map.mathshell.org/download.php?fileid=1750. That lesson demonstrates fully how students can be "invited in" to exploring the riches hinted at in the discussion above. It has ideas you can use, even if you decide that you only have time to use a small percentage of them this time.

3   The reason we talk about "bringing richer raw materials into the classroom" is that plans and materials only get you so far. Every class is different. How you think about tasks before instruction, and how tasks play out during a lesson, depends very much on who the students are and the classroom culture that you've constructed!

2

The targets in use
An extended example

This chapter illustrates how using targets from each of the five Teaching for Robust Understanding (TRU) dimensions can enrich a single class activity. For reasons of space, we'll limit this discussion to one target per dimension.

We start with a practice sheet similar to those found in many 8th grade or Algebra 1 texts and online resources – see Figure 2.1. This kind of worksheet is fine as a set of practice problems, and there's certainly a role for those. The

---

Solving Word Problems by Substitution

Instructions: For each problem, write a system of linear equations to model the situation and solve the system using the substitution method.

1. Alice painted her room using b cans of blue paint and p cans of purple paint. She used exactly 100 ounces of paint in total. Alice used 2 more cans of purple paint than blue paint. Blue paint comes in 19 ounce cans and purple paint comes in 4 ounce cans. How many cans of each color paint did she use?

2. You run the concession stand at your school's baseball game. On Friday night, you sold h hot dogs and s sodas. You sold 10 more sodas than hot dogs. Sodas cost $1.50 and hot dogs cost $2.50. In total, you made $115. How many hot dogs and sodas did you sell?

Etc....

Figure 2.1 A typical word problem worksheet.

DOI: 10.4324/9781003376903-2

second problem is included to provide a sense of the original activity, but our focus here is on enhancing the first problem.

Dimension 1: The Mathematics

We suggest you begin by solving both of the tasks as written. What mathematical issues do you see arising that might be worth highlighting for your students? What do you want them to learn?

In addition to practicing solution methods such as substitution, we see "systems of linear equations" as a mathematically rich topic that offers room for students to become increasingly comfortable mathematizing real-world situations, choosing and applying various solution strategies, and making connections between multiple representations, including graphs, tables, equations, and problem contexts. That is, there's room to do more with these tasks.

The three mathematics targets are given toward the end of Chapter 3. We'll use the second mathematics target, "In what ways do students engage with the mathematical content? What connections are built between procedures, underlying concepts, and meaningful contexts of application?" to look for enrichments to the paint problem.

Where do you think the worksheet "lands" on the target, and what opportunities are there for enrichment? See Figure 2.2 for our view.

As we see it, the worksheet lands mostly on the outer ring of this target: students repeat the same approach to a series of somewhat contrived word problems. These challenges are highlighted with dashed ovals. Fortunately, the inner rings of the target suggest some ways to open up the task. These opportunities, highlighted with solid ovals, suggest we could enhance this task by inviting students to try to solve one of these problems using a wide variety of strategies and representations, and then have the students compare and contrast the different methods.[1]

One way to support students in more authentic explorations of different strategies is to pose open-ended tasks early in the unit, *before* providing formal instruction in strategies like the substitution method. Problem 1 on the worksheet is promising as a unit entry task because of the small numbers involved. Students could solve this problem using guess and check, or tables or diagrams, and then make connections between these initial methods and more sophisticated methods such as graphing or equations. Figure 2.3 provides a version of problem 1 rewritten as a multiple strategies task for group work.[2]

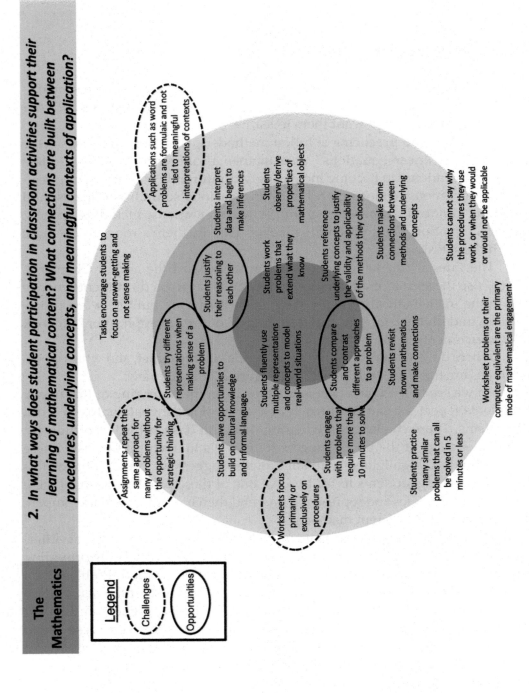

The Mathematics

2. *In what ways does student participation in classroom activities support their learning of mathematical content? What connections are built between procedures, underlying concepts, and meaningful contexts of application?*

Applications such as word problems are formulaic and not tied to meaningful interpretations of contexts

Tasks encourage students to focus on answer-getting and not sense making

Students interpret data and begin to make inferences

Students observe/derive properties of mathematical objects

Students justify their reasoning to each other

Students work problems that extend what they know

Students reference underlying concepts to justify the validity and applicability of the methods they choose

Students make some connections between methods and underlying concepts

Students cannot say why the procedures they use work, or when they would or would not be applicable

Students try different representations when making sense of a problem

Students fluently use multiple representations and concepts to model real-world situations

Students compare and contrast different approaches to a problem

Students revisit known mathematics and make connections

Students have opportunities to build on cultural knowledge and informal language.

Students engage with problems that require more than 10 minutes to solve

Worksheet problems or their computer equivalent are the primary mode of mathematical engagement

Assignments repeat the same approach for many problems without the opportunity for strategic thinking

Worksheets focus primarily or exclusively on procedures

Students practice many similar problems that can all be solved in 5 minutes or less

Legend

Challenges

Opportunities

Figure 2.2 Challenges and opportunities on the second mathematics target.

The situation:

Ms. Langill just repainted her classroom, using her own mixture of blue and purple paint to make the perfect color. Ms. Sayavedra saw the room and loved it! She wants to paint her own classroom exactly the same color.

Unfortunately, Ms. Langill can't remember exactly how many cans of paint she mixed together. She remembers that she needed 100 ounces of paint, and that there were 2 more cans of purple paint than blue paint. The blue paint came in 19 ounce cans and the purple paint, because it was snazzy and more expensive, came in 4 ounce cans.

Your job:

With your team, create a poster showing as many ways as you can to figure out how much blue paint and purple paint Ms. Langill used. Be sure that your poster shows clearly what you did and why you did it.

Figure 2.3   A modified version of the first worksheet problem.

Dimension 2: Cognitive Demand

Posing a challenging task to students prior to formal instruction on strategies can feel scary. We might wonder, "What if my students just give up?" or, at the other extreme, "What if my students *do* persevere for a long time, but still don't make progress, and lose motivation to try challenging problems in the future?" These are questions of cognitive demand. Let's take a look at target 2 from Dimension 2, Figure 2.4.

To address these challenges we can think about classroom norms that can be reinforced or created to make sure the classroom is a safe space for students to take intellectual risks. Then we can anticipate in advance what ideas students may come up with and what questions might advance their thinking without guiding them too strongly into a particular approach.

Here are some questions that can sustain a high level of cognitive demand through task completion.

- Can you explain the situation in your own words?
- What are you trying to find?
- How could you represent that?
- I see (group member) is writing something. Can you share it with the group?
- Both (group member) and (group member) have interesting ideas. Can you talk to each other about how they are similar and how they are different?
- Have you used all the important information?
- Would it help to try making diagrams, tables, equations, or graphs?
- Can you use color, arrows, or labels to show connections between your methods?

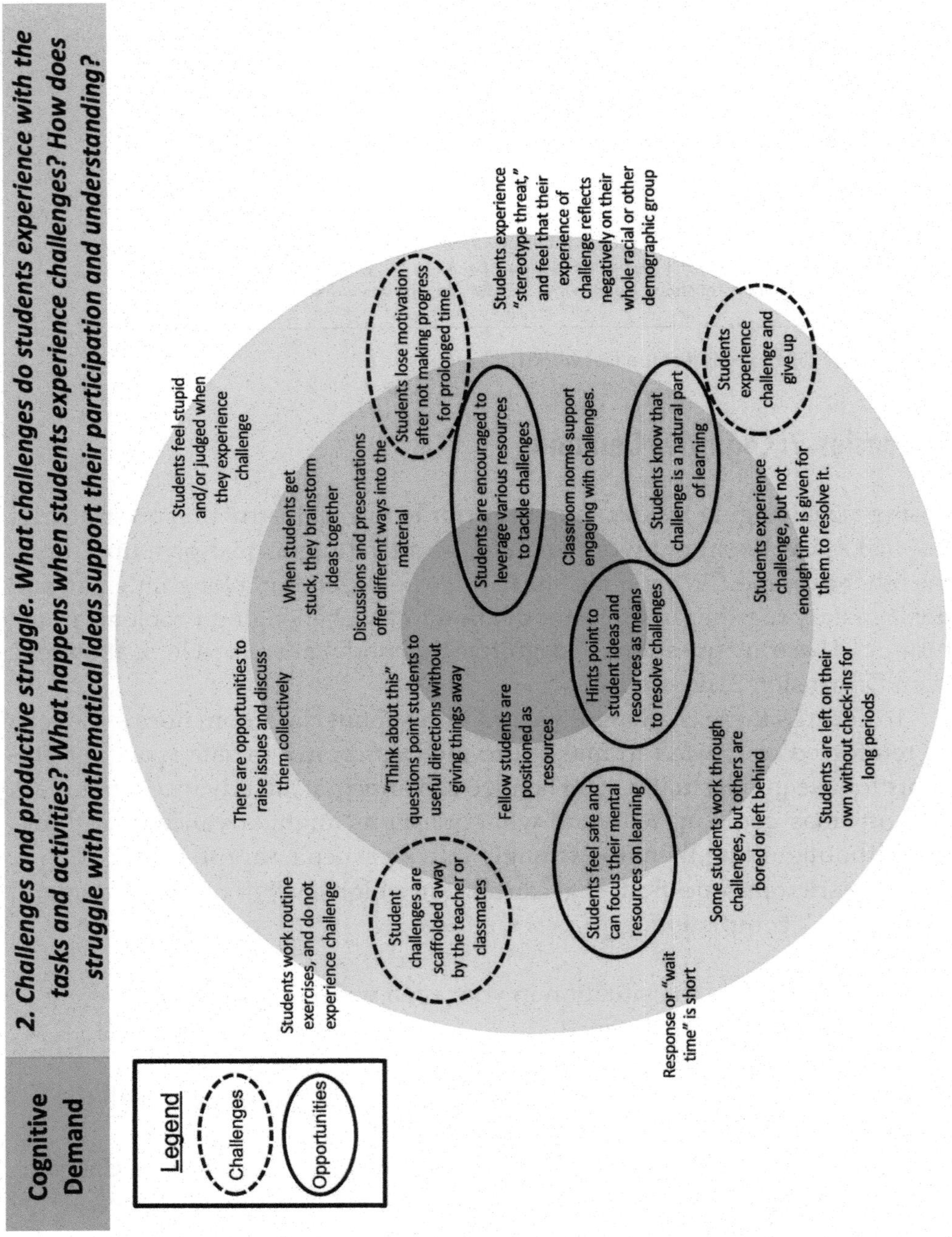

Cognitive Demand

2. Challenges and productive struggle. What challenges do students experience with the tasks and activities? What happens when students experience challenges? How does struggle with mathematical ideas support their participation and understanding?

Legend

- - - Challenges

—— Opportunities

Students experience "stereotype threat," and feel that their experience of challenge reflects negatively on their whole racial or other demographic group

Students feel stupid and/or judged when they experience challenge

Students lose motivation after not making progress for prolonged time

When students get stuck, they brainstorm ideas together

Students experience challenge and give up

Discussions and presentations offer different ways into the material

Students are encouraged to leverage various resources to tackle challenges

Students know that challenge is a natural part of learning

There are opportunities to raise issues and discuss them collectively

Classroom norms support engaging with challenges.

"Think about this" questions point students to useful directions without giving things away

Hints point to student ideas and resources as means to resolve challenges

Students experience challenge, but not enough time is given for them to resolve it.

Fellow students are positioned as resources

Student challenges are scaffolded away by the teacher or classmates

Students feel safe and can focus their mental resources on learning

Some students work through challenges, but others are bored or left behind

Students are left on their own without check-ins for long periods

Students work routine exercises, and do not experience challenge

Response or "wait time" is short

Figure 2.4 Challenges and opportunities on the second cognitive demand target.

How students respond to such questions and which students get to engage in productive struggle depend to a large degree on classroom norms, a key factor in equitable access. See the next section.

Dimension 3: Equitable Access

The revised version of the task in Figure 2.3 is mathematically rich and has room for student sense making. The question with regard to equitable access is: who gets to participate, and in what ways? Access is equitable when every student gets to engage meaningfully with core mathematical content and practices. Challenges and opportunities are highlighted in Figure 2.5.

Figure 2.5 indicates that our work in Dimension 1 has enhanced the opportunities for students to engage with important mathematical ideas and practices in meaningful ways. The central task in the modified activity is:

With your team, create a poster showing as many ways as you can to figure out how much blue paint and purple paint Ms. Langill used. Be sure that your poster shows clearly what you did and why you did it.

There are many ways to approach this version of the problem, making it potentially more accessible to more students than the original version. Requiring explanations can engage more students in that central mathematical practice. Putting the poster together provides opportunities for discussions, writing, and listening. In consequence, the revised task has increased potential to support equitable access.

Whether it does or not will depend in significant ways on the norms that have been established for group work. If productive group dynamics have not been established, for example, some students may dominate group discussions, shutting out other students. But if productive group norms exist, then students may well support each other's learning through questioning, listening, and sharing ideas.

One way to develop such norms is to provide students with sentence starters that promote mathematical thinking and respectful discourse. Examples, such as[3]:

To explain: "The strategy I used was…"; "I noticed that…"
To Clarify: "Can you explain how/why…"; I have a question about…?
To Disagree: "I disagree with _____ because…"

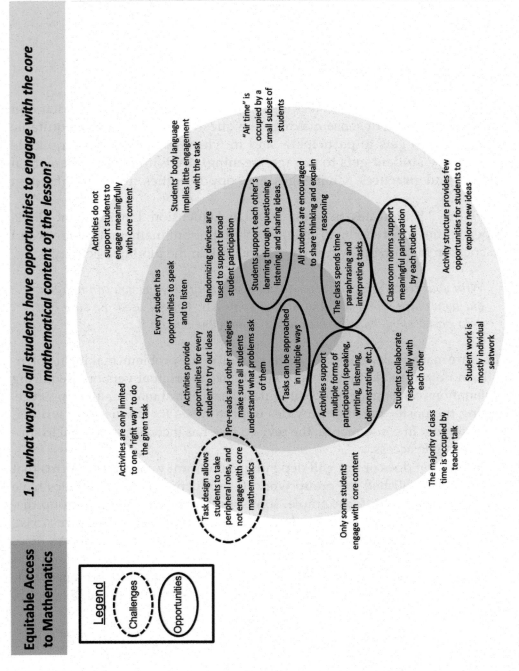

Figure 2.5 Challenges and opportunities on the first equitable access target.

Facilitators: Help your group get a quick start and keep trying ideas:
"*Can someone summarize what we are trying to do?*"
"*What could we try next?*"

Recorder/Reporters: Help your team find ways to organize your thinking for your poster.
"*How can we show this?*"
"*What should we write?*"

Resource Managers: Help your team decide when it's time to ask a team question.
"*Does anyone have any more ideas?*"
"*Are we stuck?*"

Task Managers: Encourage all teammates to explain their thinking.
"*How do you know?*"
"*Sara, what are you thinking?*"

Figure 2.6 Some classroom roles and questions that facilitate access.

can help students interact productively and invite each other into conversations. More structured approaches such as Complex Instruction (Cohen & Lotan 2014) assign students roles that support equitable group dynamics. People in each role (in boldface in Figure 2.6) can be provided with conversation prompts like those shown in Figure 2.6.

Such questions support all students, especially emerging bilingual students and students with learning differences, by providing concrete phrases that can be used to engage in the mathematical practices intended for the task. Introducing and using such conversational tools can support more equitable interactions, but inequities can still persist. Are certain students, or groups of students, systematically relegated to low-status or non-mathematical activities or shut out of conversations? Monitoring and addressing such issues is a matter of formative assessment, with significant implications for agency, ownership, and identity.

Dimension 4: Agency, Ownership, and Identity – AOI for short

The situation for agency, ownership, and identity (AOI) is similar to that for equitable access. Task design can open things up for positive identity development. Then what matters is how things play out in classroom interactions. We'll look at AOI target 1. See Figure 2.7.

The task as modified to this point already includes substantial opportunities for students' mathematical identity development, reflected in one item in the innermost ring: "Activities allow students to see themselves as problem solvers." Thus, the task adaptations suggested by the targets from

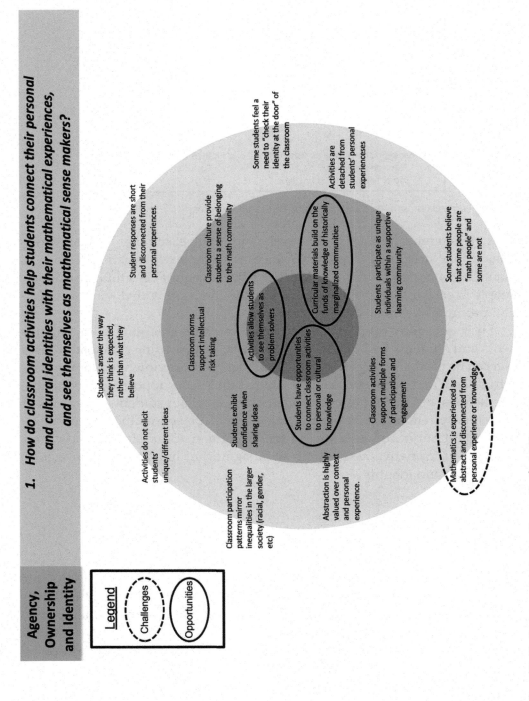

Agency, Ownership and Identity

1. How do classroom activities help students connect their personal and cultural identities with their mathematical experiences, and see themselves as mathematical sense makers?

Legend

‑ ‑ ‑ Challenges

⎯ Opportunities

Student responses are short and disconnected from their personal experiences.

Students answer the way they think is expected, rather than what they believe

Activities do not elicit students' unique/different ideas

Classroom participation patterns mirror inequalities in the larger society (racial, gender, etc)

Students exhibit confidence when sharing ideas

Classroom norms support intellectual risk taking

Abstraction is highly valued over context and personal experience.

Activities allow students to see themselves as problem solvers

Students have opportunities to connect classroom activities to personal or cultural knowledge

Classroom culture provide students a sense of belonging to the math community

Curricular materials build on the funds of knowledge of historically marginalized communities

Students participate as unique individuals within a supportive learning community

Classroom activities support multiple forms of participation and engagement

Some students feel a need to "check their identity at the door" of the classroom

Activities are detached from students' personal experienceses

Some students believe that some people are "math people" and some are not

Mathematics is experienced as abstract and disconnected from personal experience or knowledge

Figure 2.7 Challenges and opportunities on the first AOI target.

Dimensions 1–3 also support Dimension 4 to some extent. Yet, much will depend on how the task is implemented in the classroom, which depends on connecting personally with individual students and establishing productive classroom norms. We need to attend to students' affective experiences as they engage in problem-solving. Do students appear to feel a sense of belonging in the math community? Are individual and cultural differences respected and valued as part of a supportive learning community? If not, that's an issue. Are there situations where students are excluded or experience disconnects between their personal, cultural, and mathematical identities? If so, these will need attention.

The target also raises important other issues. This task offers opportunities for some aspects of AOI but does not achieve cultural relevance and connections to students' lives. Not every task can do everything, but the AOI targets point to gaps that should be addressed somewhere, if not here.

Specifically, the problem context appears contrived. The problem situation and numbers in the problem seem arbitrary, designed for ease of solving rather than seeming authentic. Who buys paint in 4 ounce or 19 ounce cans or mixes exactly 100 ounces? While the use of "easy" numbers supports the multiple strategies aspect of the task, the target suggests a missed opportunity to invite students to connect classroom activities to personal or cultural knowledge or to use a contextual problem that would build on the funds of knowledge of historically marginalized communities.

We consider the modified paint problem to be a reasonable and appropriate task. However, over the course of a unit or school year, other lessons could be included that hit the center of this target, such as lessons involving data collection, social justice, and/or community funds-of-knowledge projects. With appropriate attention to the ways students experience *doing* mathematics and the connections to their personal and cultural identities, both long projects and shorter tasks can support Agency, Ownership, and Identity.

Dimension 5: Formative Assessment

The fifth TRU dimension, Formative Assessment – bringing students' thinking out into the open and then adjusting instruction to address the challenges or possibilities that have been revealed – provides the "fine-tuning" that enhances classroom activities. Eliciting and attending closely to students' mathematical thinking during instruction is a key aspect of ensuring that students are working in rich mathematical territory (Dimension 1), in ways that involve sense making and productive struggle (Dimension 2). If a student seems discouraged, excluded, or not actively engaged, asking them about their mathematical thinking and building on some aspect of their attempted work are important ways to increase access (Dimension 3). Further, creating rich opportunities for students to share and build on each other's mathematical thinking also creates opportunities for student agency, ownership, and identity (Dimension 4).

Much of the "action" with regard to formative assessment occurs through classroom interactions. Understanding and responding to what students are thinking and doing in real time can be challenging. However, lesson preparation can make it easier to act in the moment. By designing tasks that are likely to support rich interactions *and by thinking in advance about the challenges students may encounter*, we're in a good position to help students move forward.

Let's consider what we might expect when students[4] dig into the modified task in Figure 2.3.

It's useful to solve the problem in as many ways as possible, while also thinking about places where students may encounter difficulties. For the paint problem, we can anticipate that students will pursue some of the strategies listed in Table 2.1.

Table 2.1 Possible student approaches to the paint problem

Diagrams and Tables	Equations	Graphs
Drawing three or four cans of blue paint (both good guesses) and two more cans of purple paint Guess and check tables with three columns: cans of blue paint (a guess), cans of purple paint (two more than the previous column), and the total ounces of paint Diagram representing the 100 ounces, for example as a 10×10 grid, and successively shading in some squares blue and some purple using the other constraints until reaching 100.	System of equations: $b + 2 = p$ $4b + 19p = 100$ or $p = b + 2$ and $4b + 19p = 100$ or $b = p - 2$ and $4b + 19p = 100$ Single equation: $4b + 19(b + 2) = 100$ or $4(p - 2) + 19p = 100$ Equivalent versions for graphing, such as $b = (-19/4)p + 25$	The pairs of equations or single equations in the "equations" column can all be graphed, allowing for visual solutions.

Specifically, here are some challenges students may encounter. You may well list more.

- Will students' drawings or tables be organized in ways that facilitate guess and check?
- Will students be confused by the use of b and p for the *number* of blue and purple cans, respectively?
- Will they be stymied by the fact that there are equivalent forms of the equations?
- What issues (either direct or strategic) might they run into when solving the equations?
- Will they think to transform equations for purposes of graphing? (Students may have some difficulty graphing $4b + 19p = 100$. Some possible strategies are to transform the equation, plot points, or find the x- and y- intercepts, but none of these strategies are straightforward.)
- How will they interpret the graphical solutions when there is one equation, or two?

All these challenges arise before students begin to compare and contrast their strategies or put their posters together! Those processes may introduce additional challenges.

It's useful to think about how to spark mathematically rich and productive conversations as the students produce and present their posters. Let's look at the first formative assessment target, as shown in Figure 2.8. As designed, the modified task does not have any apparent challenges, but we think there still exist some opportunities for improvement.

Opportunities in Figure 2.8 relate to the fact that there are multiple ways for students to approach and solve the problem, that students are actively encouraged to explain what they are doing, and that they are asked to compare and contrast varied approaches to the task.

It would be a missed opportunity if students try to solve this task using a "divide and conquer" approach, in which each student individually comes up with a strategy and writes it on the poster, without comparing them. One simple change can make the expectation of comparing and connecting strategies much more explicit. We might consider the addition of something like the underlined text.[5]

With your team, create a poster showing as many ways as you can to figure out how much blue paint and purple paint Ms. Langill used. Be sure that your poster shows clearly what you did and why you did it. <u>Use color, arrows, labels, and/or written explanations to show connections between strategies.</u>

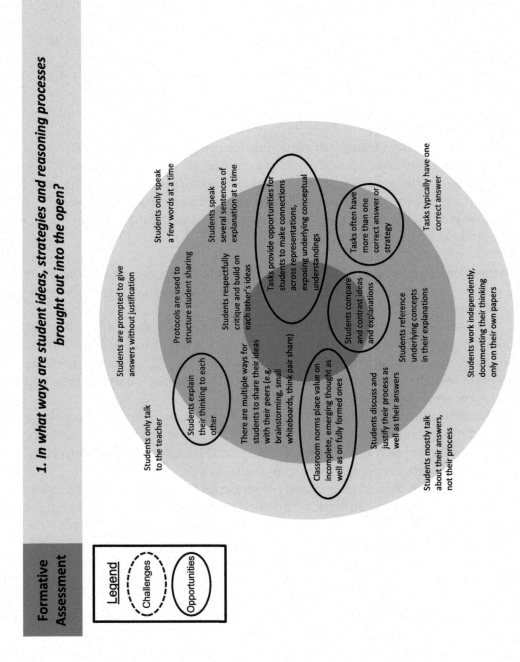

Formative Assessment

1. In what ways are student ideas, strategies and reasoning processes brought out into the open?

Students only speak a few words at a time

Students speak several sentences of explanation at a time

Students are prompted to give answers without justification

Protocols are used to structure student sharing

Students respectfully critique and build on each other's ideas

Tasks provide opportunities for students to make connections across representations, exposing underlying conceptual understandings

Tasks often have more than one correct answer or strategy

Tasks typically have one correct answer

Students only talk to the teacher

Students explain their thinking to each other

There are multiple ways for students to share their ideas with their peers (e.g. brainstorming, small whiteboards, think pair share)

Classroom norms place value on incomplete, emerging thought as well as on fully formed ones

Students compare and contrast ideas and explanations

Students reference underlying concepts in their explanations

Students discuss and justify their process as well as their answers

Students work independently, documenting their thinking only on their own papers

Students mostly talk about their answers, not their process

Legend

Challenges

Opportunities

Figure 2.8 Opportunities on the first formative assessment target.

This additional sentence implies that students should not only discuss their strategies with each other but also discuss connections and contrasts and highlight these on the poster. It calls for collaborative decision-making both on mathematical substance and on poster preparation. (For example, if students choose to use a blue color throughout the poster for cans of blue paint and a purple color throughout the poster for cans of purple paint, they will likely need to discuss in some detail where each color should be used before putting pen to paper.)

Finally, how this task plays out in instruction depends once again on classroom norms. Are students comfortable exchanging ideas and working collaboratively? Have they learned to value incomplete thoughts and work to refine them? The right norms and support structures can help to achieve the potential of this task as highlighted in the five targets discussed in this chapter.

Postscript

That was a lot, we know. It's like any sport or any complex activity when you first take it up. The complexity is almost overwhelming at first, because you have to stop and think everything through. But once you get the hang of it, it becomes second nature. The first major goal is *to learn to think with TRU.* That comes with practice. Then, when we look at any proposed task or activity, the big TRU questions come naturally: How can we make the math richer? Are there handholds for students, so they can engage in productive struggle? Who participates, in what ways? Once you have that perspective, the targets serve as *reminders.* You can scan a target quickly and say, "this is a challenge with this task and here's an opportunity that I can take advantage of." Thousands of teachers will tell you that it doesn't take that long to get there and that the effort pays off.

Notes

1   In addition to enriching the task, opening it up to multiple approaches invites more students into the mathematical conversation and provides opportunities for conversations that can develop students' agency, ownership, and identity – TRU Dimensions 3 and 4 respectively.

2   This version of the task was written by Johanna Langill and Alyssa Sayavedra, Oakland Unified School District, California.

3   See https://mrswintersbliss.com/product/math-talk-sentence-starters-%E2%80%A2-math-talk-posters/.

4   We're writing about generic "students." When you're thinking about your own students, you can be more specific.

5   We credit Ms. Langill in the Oakland Unified School District for this idea.

3

Dimension 1
The Mathematics

What the mathematics dimension involves

We want our students to become knowledgeable, flexible, resourceful, and agentive mathematical thinkers and problem-solvers. The tasks and activities they engage with in our classrooms should provide them the opportunities to grow in these directions.

In any lesson, the quality of the mathematics discussed shapes what students are likely to learn. Here are the central questions we ask.

- What are the essential mathematical ideas students will grapple with in the lesson?
- What mathematical practices, ways of thinking, and habits of mind will they use?
- How is what they're supposed to learn connected to what they know, both in and out of the classroom?
- How is it connected to what they've learned before and will learn later?

It's common to think about lessons in terms of the skills that students will learn or practice. Skills are important, but when we talk about "the mathematics" in TRU terms, we think beyond skills. We think about the concepts that underlie the skills, the mathematical practices and habits of mind students will use, and the connections between ideas. These concepts, practices, and connections are what tie together individual lessons and activities across the year. They are what allow students to see and experience mathematics as a coherent, connected, sense-making discipline.

DOI: 10.4324/9781003376903-3

A bit more background

Following a decade of "back to basics," the National Council of Teachers of Mathematics recommended in 1980 that "problem solving be the focus of school mathematics in the 1980s" (NCTM, 1980, p. 1). Problem-solving and, more generally, mathematical thinking soon became a major focus of research and practice (see, e.g., Schoenfeld, 1985, 1992). In 1989, NCTM issued the *Curriculum and Evaluation Standards for School Mathematics*. For the first time curricular goals included not only mathematical content (number, algebra, geometry, statistics and probability, etc.) but key mathematical processes: mathematics as problem-solving, mathematics as communication, mathematics as reasoning, and making mathematical connections. The field's sense of what it is to *do* mathematics expanded significantly.

NCTM's (2000) *Principles and Standards* updated the content and added the idea of using mathematical representations as a key practice. A decade later, the *Common Core State Standards for Mathematics* (CCSSI, 2010) provided another update, in much the same spirit. Central content and practices (e.g., problem-solving, reasoning, making mathematical arguments, modeling, and using appropriate tools strategically) are intended to be at the heart of mathematics instruction, as specified in the Common Core documents and the vast majority of state standards. Here we highlight a few key points.

Mathematical content can and should be connected and coherent, and it should make sense to students.

Nothing our students encounter – whether it is how to combine or reduce fractions, how to solve quadratic equations by factoring or using the quadratic formula, how to find solutions of families of equations, or how to solve complex modeling problems – is arbitrary. If students memorize procedures, they are likely to forget them. If they understand where procedures come from, they can re-generate them.

Wertheimer (1945) wrote about a class of students who worked the following task:

$$\frac{274+274+274+274}{4}=$$

The vast majority of students first added the four 274s to get 1096 and then divided 1096 by 4 to get 274. In doing so they missed the point of the task: if you multiply something by 4 (which is what you do when you add something four times) and then divide by 4, you end up where you started. Yes, we want students to be able to perform the basic operations – *and* we want them to understand what they're doing.

This example contrasts with what happened when a retired mathematician was asked how he would solve a classic word problem of the type "Pipe A can fill a pool in 3 hours. Pipe B can fill a pool in 4 hours. How long does it take to fill the pool if both Pipe A and Pipe B are being used? He said, "I haven't seen a problem like this in 40 years. There's a formula, but there's no way I can remember it." Then he said, "OK, let's see. What can I combine? Not hours, not pools … but I can add rates. Pipe A fills ⅓ of the pool in an hour, and pipe B fills ¼ of the pool in an hour, so together they fill (⅓ + ¼) = 7/12 of the pool in 1 hour. It takes 12/7 hours to fill the pool."

The mathematician didn't depend on memorized formulas, although if he'd remembered the formula he'd have used it. His solution demonstrates the kind of thinking we'd like to see from our students. It's important to teach in ways that help students see connections and develop sense-making habits. This next subsection provides an example.

There are many ways to "see" or represent mathematical objects. Having a variety of representations at their disposal helps students become more powerful thinkers.

Take a moment to think about this classic problem:

> *Train A leaves a train station at noon and travels at a steady speed of 50 miles per hour. Three hours later Train B leaves the station on a parallel track, traveling at a steady speed of 60 miles per hour. How long does it take for Train B to catch up with Train A?*

If you ask college students to solve this problem in any way they feel comfortable, many of them will build tables; some will draw graphs and some will use algebra. When teaching we typically move to the most general and powerful solution, which is algebraic. But, in interesting ways, the algebra hides what is going on. When they make tables, students can see how train B is getting closer, over time. (Moreover, when they make tables on their own, that may be because tables are the math they're most comfortable with.) Can you see Train B catching up with Train A in the graphical solution? Can you see it in the algebraic solution?

Providing students opportunities to represent mathematical phenomena in different ways helps them to make connections and see mathematics as a coherent whole. It also provides more students different "ways into" the math, so it serves as a way of increasing equitable access to the content. We've asked groups of college students to make posters showing as many solutions as possible and to explain their work on the posters. Then we discuss their posters as a whole class. Those who started with tables see the links to graphs and algebra, and those who started with the algebra are challenged to see where "catching up" happens in the algebraic solution. That kind of approach allows all students to engage meaningfully with the core content. It enriches mathematics, offers a range of cognitive demand so that more students are "stretched," and promotes equitable access.

Students need opportunities to develop productive mathematical dispositions and habits of mind.

Students won't get good at mathematical reasoning and problem-solving (or any of the other practices) unless they have ample opportunity to engage in them. Along these lines, it is worth keeping in mind the "five intertwined strands of mathematical proficiency" described by the National Research Council (2001, p. 5): productive disposition, conceptual understanding, procedural fluency, strategic competence, and adaptive reasoning.

When we think about the mathematics involved in a lesson, we think about opportunities for students to develop these aspects of proficiency. Where do students have opportunities to explore? To conjecture? To test their ideas, to refine their thinking? The more that students have opportunities to *do* mathematics (with support and encouragement), the more they are likely to develop productive dispositions and habits of mind in addition to core knowledge, skills, and practices.

It's essential to make connections not only within mathematics but also between mathematics and the "real world." It's important in mathematical terms because mathematical modeling is a powerful way of making sense of the world around us. It's comparably important in personal terms. For many students, school mathematics has little or nothing to do with their personal lives – so why study it? Yet, there are so many ways in which mathematics can and does relate to students' lived experiences. Engaging with such mathematics expands students' mathematical understanding; it can contribute to their sense of agency and ownership of the mathematics and their mathematical identities. See Chapter 6 regarding AOI, specifically the references to culturally relevant pedagogy and to curricular examples that build bridges to and from students' experiences.

Some ideas that may help

There are many ways to open up the mathematics in a lesson. As illustrated in the introduction, problems can be re-framed so that students have opportunities to do more of the mathematical work themselves – e.g., students can conjecture results and then prove or disprove them, rather than being instructed to prove things they are told are true.

Almost all problems worth thinking through can be posed such that multiple approaches and strategies are possible. If students approach a task in different ways, asking them to compare, contrast, and connect the ideas that they produce deepens their mathematical experiences. Multiple approaches provide ways to engage with the content at different levels of cognitive demand, providing access to mathematics to more students. (See, e.g., Lotan, 2003, who discusses "group worthy" problems.) As we noted in the discussion of the "catching up" train problem, making connections between different approaches can help students develop a deeper understanding of central mathematical ideas.

Asking for justifications or applications provides students with opportunities to delve into the mathematical practices. A number of questions are often applicable and can be used to nudge students more deeply into mathematics. For example:

- Is that always true? How do we know?
- Can you think of another way to solve this problem?
- How are the two solutions methods related to each other?
- Are there reasons why you may want to use one method over the other?
- How would you describe your strategy to someone who is two grades below you?
- What do you know about this topic? What problem-solving strategies might be helpful here?

Making a habit of expanding mathematical opportunities enhances students' experience with mathematics as a deeply connected discipline and sets the stage for their developing the full range of mathematical proficiencies described above.

More specific questions for planning and review

Here are some specific questions you might consider as you reflect on the opportunities for mathematical richness provided by a lesson plan or reflect on how a lesson went. Some are drawn from the *TRU Math Conversation Guide* (Baldinger, Louie, and the Algebra Teaching Study and Mathematics Assessment Project, 2018).

- What are the mathematical goals for the lesson?
- What connections exist (or could exist) between important ideas in this lesson and important ideas in past and future lessons?

- How does this lesson draw on students' outside-of-school knowledge and language repertoires?
- How do important mathematical practices develop in this lesson/unit?
- How are facts and procedures in the lesson justified?
- How are facts and procedures in the lesson connected with important ideas and practices?
- What opportunities do students have to do problem-solving, work strategically, and engage in mathematical reasoning and other central mathematical practices? Are there ways to open up tasks in the curriculum or lesson plan?
- What opportunities do students have to practice using mathematical language, not only key vocabulary but also using language to make sense of problems, suggest approaches, justify their thinking, and constructively critique others' ideas?
- How can students be given opportunities to extend their ideas, use different representations, or generalize patterns?

Some resources that might be useful

Appendix A contains a collection of classroom strategies that serve multiple purposes. Specific strategies that are often useful for expanding the mathematics in a task or activity include: *What's the big picture?*; *Always, Sometimes, Never*; *Ask for patterns and outliers*; *Deep dive homework*; *Invent an argument*; *Make categories*; *Open it up*; *Reflective Journals*; *Student-led questioning routines*; and *What do you wish you knew?*

The SERP "5 by 8 card" at https://www.serpinstitute.org/5x8-card and "the deck behind the card" at https://serpmedia.org/deck-behind-5x8-card/ provides links to a number of useful techniques – similarly, for multiple purposes.

NCTM offers a wide range of resources at http://www.nctm.org/. NCTM's statewide affiliates have annual meetings that are informative; other groups such as Teachers' math circles consistently dig into rich mathematics. The following article and video series might be of interest.

- This article on habits of mind by Al Cuoco and others expands habits of mind specifically for high school algebra and geometry: https://nrich. maths.org/content/id/9968/Cuoco_etal-1996.pdf
- This video series from NCTM on teaching with the CCSS: http://www. nctm.org/Standards-and-Positions/Common-Core-State-Standards/ Teaching-and-Learning-Mathematics-with-the-Common-Core/

The mathematics targets

Here is our first set of targets. The big math question is: In what ways do classroom activities provide opportunities for students to become knowledgeable, flexible, and resourceful mathematical thinkers? This is broken down into three main sub-questions, each of which serves as the topic for one target. See Figure 3.1.

- Target 1: What is the main mathematical idea? How does it develop? How is it connected to what students know? How is it connected to the grade-level content and practice standards? (Figure 3.2).
- Target 2: In what ways does student participation in classroom activities support their learning of mathematical content? What connections are built between procedures, underlying concepts, and meaningful contexts of application? (Figure 3.3).
- Target 3: In what ways does student participation in classroom activities support the development of important mathematical practices and other productive mathematical habits of mind? (Figure 3.4).

To use the targets, think about a specific task that you plan to have students work on. Now look at Target 1, and think about where the task "lands." Then, look at the attributes that sit closer to the center of the target. Is there a way to modify the task so that it lands on some of those? Then do the same for targets 2 and 3 – subject to the time and other constraints you're working with. The result should offer students a richer and deeper mathematical experience.

We can't do this for every task – and sometimes, "less is more." But if we get in the habit of thinking about how our students can experience tasks in ways that offer deeper engagement with mathematical content and practices, their experience of mathematics will be richer. Picking one activity for each lesson and thinking about enhancing it would be a great start.

After you teach the enhanced lesson, take some time to think about how the activities played out in practice. Did some things go better than you expected? If so, what did you do to help good things happen? If some things didn't go well as planned, what might you do differently next time?

In addition to the three filled-in targets in this chapter, we offer a final blank target (Figure 3.5). You may want to articulate your own goals or ideas on that target. We hope the targets will be useful to you in developing opportunities for your students to become knowledgeable, flexible, and resourceful mathematical thinkers.

The Mathematics

In what ways do classroom activities provide opportunities for students to become knowledgeable, flexible, and resourceful mathematical thinkers?

Core Questions:

1. **What is the main mathematical idea?** How does it develop? How is it connected to what students know? How is it connected to the grade level content and practice standards?

2. **In what ways does student participation in classroom activities support their learning of mathematical content?** What connections are built between procedures, underlying concepts, and meaningful contexts of application?

3. **In what ways does student participation in classroom activities support the development of important mathematical practices** and other productive mathematical habits of mind?

Figure 3.1 Core questions about Dimension 1, the mathematics.

The Mathematics

1. What is the main mathematical idea? How does it develop? How is it connected to what students know? How is it connected to the grade level content and practice standards?

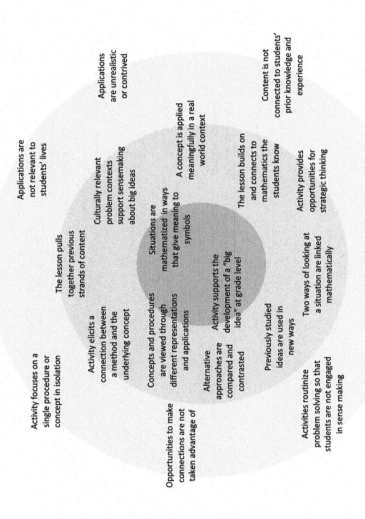

Applications are not relevant to students' lives

Applications are unrealistic or contrived

Content is not connected to students' prior knowledge and experience

The lesson pulls together previous strands of content

Culturally relevant problem contexts support sensemaking about big ideas

A concept is applied meaningfully in a real world context

Situations are mathematized in ways that give meaning to symbols

The lesson builds on and connects to mathematics the students know

Activity provides opportunities for strategic thinking

Activity focuses on a single procedure or concept in isolation

Activity elicits a connection between a method and the underlying concept

Activity supports the development of a "big idea" at grade level

Concepts and procedures are viewed through different representations and applications

Two ways of looking at a situation are linked mathematically

The content is below grade level in ways that do not support grade level objectives

Alternative approaches are compared and contrasted

Previously studied ideas are used in new ways

Opportunities to make connections are not taken advantage of

Activities routinize problem solving so that students are not engaged in sense making

Figure 3.2 Mathematics Target 1, what is the main mathematical idea?

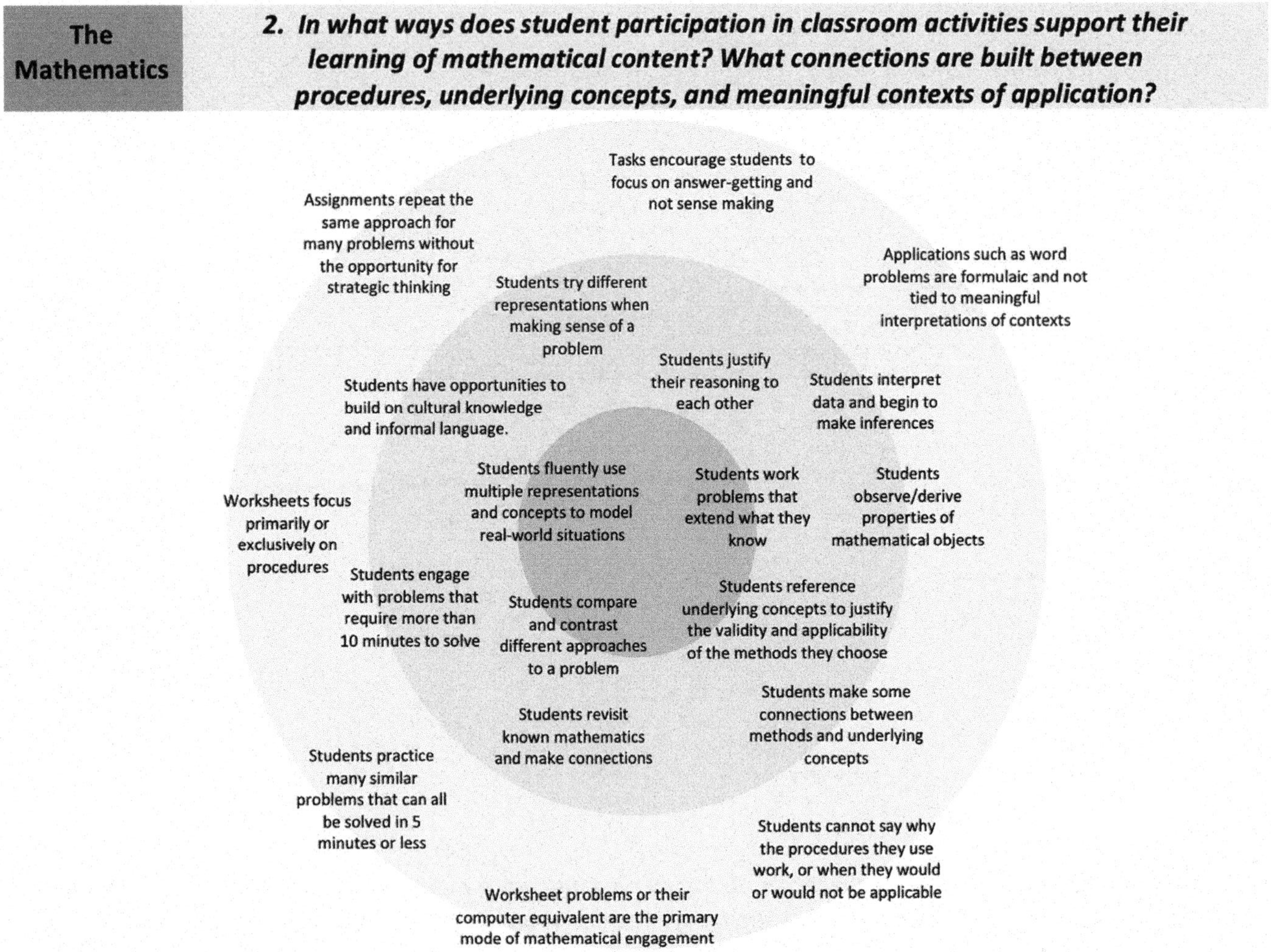

The Mathematics

2. In what ways does student participation in classroom activities support their learning of mathematical content? What connections are built between procedures, underlying concepts, and meaningful contexts of application?

Tasks encourage students to focus on answer-getting and not sense making

Applications such as word problems are formulaic and not tied to meaningful interpretations of contexts

Students interpret data and begin to make inferences

Students observe/derive properties of mathematical objects

Students try different representations when making sense of a problem

Students justify their reasoning to each other

Students work problems that extend what they know

Students reference underlying concepts to justify the validity and applicability of the methods they choose

Students make some connections between methods and underlying concepts

Assignments repeat the same approach for many problems without the opportunity for strategic thinking

Students fluently use multiple representations and concepts to model real-world situations

Students compare and contrast different approaches to a problem

Students revisit known mathematics and make connections

Students cannot say why the procedures they use work, or when they would or would not be applicable

Students have opportunities to build on cultural knowledge and informal language.

Students engage with problems that require more than 10 minutes to solve

Worksheets focus primarily or exclusively on procedures

Students practice many similar problems that can all be solved in 5 minutes or less

Worksheet problems or their computer equivalent are the primary mode of mathematical engagement

Figure 3.3 Mathematics Target 2, student engagement with mathematical ideas.

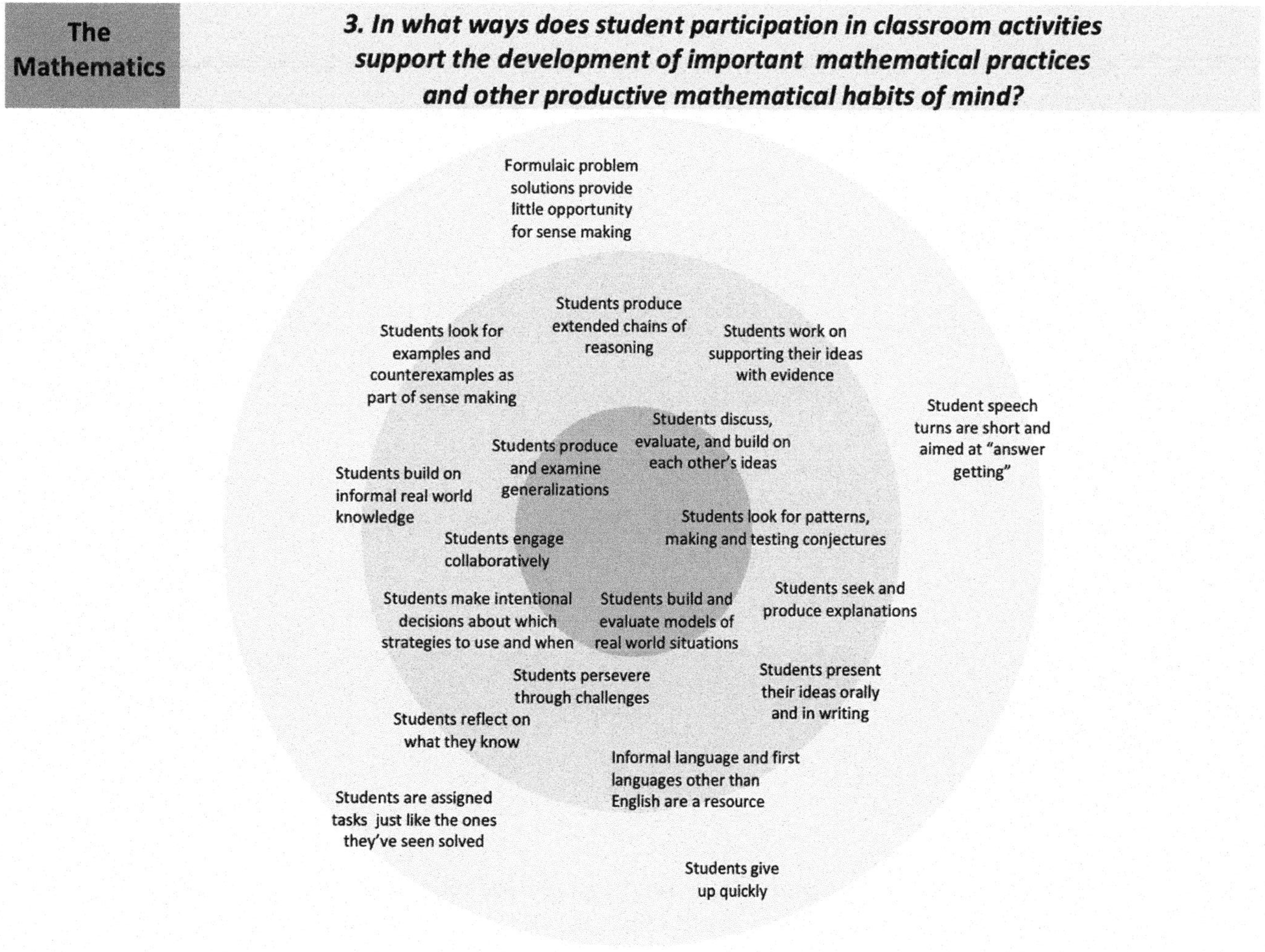

The Mathematics

3. In what ways does student participation in classroom activities support the development of important mathematical practices and other productive mathematical habits of mind?

Student speech turns are short and aimed at "answer getting"

Students work on supporting their ideas with evidence

Students look for patterns, making and testing conjectures

Students seek and produce explanations

Students present their ideas orally and in writing

Formulaic problem solutions provide little opportunity for sense making

Students produce extended chains of reasoning

Students discuss, evaluate, and build on each other's ideas

Students produce and examine generalizations

Students build and evaluate models of real world situations

Students give up quickly

Students look for examples and counterexamples as part of sense making

Students engage collaboratively

Students make intentional decisions about which strategies to use and when

Students persevere through challenges

Informal language and first languages other than English are a resource

Students build on informal real world knowledge

Students reflect on what they know

Students are assigned tasks just like the ones they've seen solved

Figure 3.4 Mathematics Target 3, mathematical practices and habits of mind.

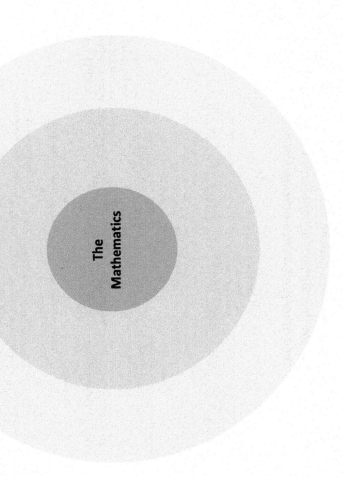

In what ways do classroom activities provide opportunities for students to become knowledgeable, flexible, and resourceful mathematical thinkers?

The Mathematics

Build your own!

The Mathematics

Figure 3.5 Mathematics Target 4, build your own target!

Using the mathematics targets – three examples

Chapter 2 provided a full-fledged example of how one target from each TRU dimension can be used to enrich a proposed activity. Here, to show how the targets can be used at all grade levels, we demonstrate how to enrich tasks at the elementary, middle, and secondary levels by considering the three mathematics targets. In the following chapters we'll do the same for the other four dimensions of TRU.

In what follows we'll identify some, but hardly all, of the relevant challenges and opportunities on each of the three mathematics targets and suggest how those observations can inspire task enhancements.

An elementary grades example – two-digit subtraction

Task enhancement strategy: name and explore students' varied methods

Imagine you're at the point in the curriculum where your students have learned two-digit subtraction. Figure 3.6 shows a typical sheet of practice examples using "borrowing."

Practicing skills is useful. But are there ways we can modify or add to the practice sheet so that students get more out of working it?

We suggest that you open this book to (or download or copy) the full-size math targets presented above and think about modifications with us. Even better, identify some possible modifications before you read our thoughts!

1) 42 − 18	2) 62 − 35	3) 55 − 28	4) 71 − 49	5) 70 − 37
6) 46 − 19	7) 73 − 26	8) 36 − 19	9) 71 − 49	10) 70 − 37
11) 68 − 39	12) 43 − 17	13) 84 − 26	14) 92 − 27	15) 80 − 22
16) 53 − 16	17) 77 − 38	18) 64 − 35	19) 74 − 58	20) 62 − 36

Figure 3.6 A typical worksheet to practice subtraction.

Identifying the challenges

Here are some limitations or challenges we see in the worksheet.

On Target 1:

- Activity focuses on a single procedure or concept in isolation.
- Activities routinize problem-solving so that students are not engaged in sense-making.
- Content is not connected to students' prior knowledge and experience.

On Target 2:

- Assignments repeat the same approach for many problems without the opportunity for strategic thinking.

On Target 3:

- Formulaic problem solutions provide little opportunity for sense-making.
- Students are assigned tasks just like the ones they've seen solved.

We certainly can't expect to address all of these challenges at once! But, the targets offer some opportunities for possible improvements.

Identifying the opportunities

On Target 1:

- Alternative approaches are compared and contrasted.
- Activity provides opportunities for strategic thinking.

On Target 2:

- Students justify their reasoning to each other.
- Students compare and contrast different approaches to a problem.

On Target 3:

- Students work on supporting their ideas with evidence.
- Students discuss, evaluate, and build on each other's ideas.

Generating possibilities

We're not trying to create a super-task that takes advantage of all these opportunities. Rather, we ask: are there ways we can take advantage of some of these opportunities to enrich the original task? You might well expand the task in different directions.

Some of the possibilities can be achieved by using number talks (see, e.g., Parrish, 2010):

- In all of these tasks the second digit of the second number is larger than the second digit of the first number, so students have to "borrow" or use whatever procedure they've learned. What about mixing things up, including some subtractions like

$$\begin{array}{r} 87 \\ -53\,? \\ \hline \end{array}$$

  Doing so helps to avoid the mindless use of the subtraction procedure. Then you can ask, "which of the problems on this sheet are easiest for you to solve, and why? What makes the others harder, and how do you think about solving them?" This opens the task up to reflection and strategic thinking.
- Ask students to explain their thinking, in either pairs or in the whole class. If they respond in procedural terms ("I cross this number out and subtract one, then add a one to this number") ask if they can explain why. Explanations open up connections between procedures and underlying concepts, provide opportunities for students to exchange ideas, and move students away from memorization and toward understanding.
- When discussing the first task,

$$\begin{array}{r} 43 \\ -19, \\ \hline \end{array}$$

  ask if anyone has solved the problem a different way. Some students may say that they changed the problem to

$$\begin{array}{r} 44 \\ -20, \\ \hline \end{array}$$

  which they thought was easier to do. This can lead to discussions of if and why this method would always work – and whether it makes sense to use this method for

$$\begin{array}{r} 79 \\ -18. \\ \hline \end{array}$$

Whether you have your students work these problems in think/pair/shares or in whole class discussion, and whether you decide to have students write up their ideas to share (perhaps through classroom jigsaws or posters) will depend on what you think your students will profit most from, the amount of time you have for discussion, and what activities the class has engaged in recently. But, all together, these kinds of activities have the potential to address every one of the opportunities highlighted above.

A middle grades example – graphing lines

Task enhancement strategy: encourage comparison of multiple solution strategies

Imagine that you're at the point in the middle grades curriculum where your students have studied different forms that represent the graphs of straight lines – e.g., the point-slope formula, standard form, slope-intercept form, two-point form, and two-intercept form.

Typical exercises call for sketching the graphs of a collection of equations that are given in the various forms, for example:

Sketch the graphs of the following lines:

$$y = \frac{3}{2}x + 7$$

$$3x - 2y = -14$$

The line that passes through (4, 13) and has slope 1.5

The line that has intercepts (0, 7) and (−14/3)

Before starting revisions, let's think about what we really want students to know. For example,

- We'd like students to be able to produce the equation of a line, given the graph – and we'd like them to be able to do so strategically. If they're given a graph that passes through a whole-number y-intercept, will they use that information? If the graph passes through two points with whole-number coordinates, will they take advantage of that fact?
- Do they know that any two pieces of information determine a line and that any line that is not parallel to either axis can be written in every single form[1,2]?

With these issues in mind, let's turn to the mathematics targets.

Identifying the challenges

On Target 1:

- Activities routinize problem-solving so that students are not engaged in sense-making.

- Content is not connected to students' prior knowledge and experience.
- Opportunities to make connections (e.g., between forms and to represent real-world phenomena) are not taken advantage of.

On Target 2:

- Students repeat the same approach for many problems without the opportunity for strategic thinking.
- Students practice many similar problems that can all be solved in 5 minutes or less.

On Target 3:

- Students work on tasks just like the ones they've seen solved.

Identisfying the opportunities

On Target 1:

- Activity elicits a connection between a method and the underlying concept.
- Two ways of looking at a situation are linked mathematically.
- Activity provides opportunities for strategic thinking.

On Target 2:

- Students compare and contrast alternative methods.
- Students reference underlying concepts to justify the validity and applicability of the methods they choose.
- Students justify their reasoning to each other.

On Target 3:

- Students work strategically.
- Students are working on supporting their ideas with evidence.

Generating possibilities

There's a large space of possibilities here! But, given the goals we outlined above and some of the opportunities highlighted in the targets, we might consider the following:

- Giving the students a linear graph and asking them in groups to determine the equation of the graph; asking if there is more than one solution, and having the students compare and contrast solutions; and having the students argue about which approach they find "easiest" or "more efficient." Then, is that always the case? What about a different graph?

(Generally speaking, when a problem can be approached in more than one way, asking students to compare and contrast their approaches can open up very interesting conversations.)

– Asking students to determine if three points are on the same straight line and to justify their method. This requires looking closely at slope and explores the reason that the point-slope formula works. (See https://www.map.mathshell.org/lessons.php?unit=8215&collection=8. This sample lesson addresses many of the opportunities highlighted above.)

A secondary grades example – exponential decay

Task enhancement strategy: choosing problem contexts that leverage students' experiences

Here is a typical exponential decay problem:

> *Over the past few years, the number of students enrolled in after-school programs has been decreasing. Each year there is a 11% decrease in student enrollment. Currently, 13,145 students are enrolled. If this trend continues, how many students will be enrolled in 6 years?*

The solution offered is a purely mechanical plug-in to the formula, $FV = PV(1 - d)^n$, where FV = future value, PV = present value, d = rate of decay, and n = number of periods. There are many exercises of this type. What can we do to enrich them?

Identifying the challenges

On Target 1:

– Application is not relevant to students' lives.
– Application is unrealistic or contrived.
– Content is not connected to students' prior knowledge and experience.

On Target 2:

– Applications such as word problems are formulaic and not tied to making sense of the contexts.

On Target 3:

– Formulaic problem solutions provide little opportunity for sense-making.
– Students work on tasks just like the ones they've seen solved.

Identifying the opportunities

On Target 1:

– The lesson pulls together previous strands of content.
– A concept is applied in a real-world context.
– Situations are mathematized in ways that give meaning to symbols.

On Target 2:

– Students compare and contrast different approaches to a problem.
– Students work problems that extend what they know.
– Students observe/derive properties of mathematical objects.

On Target 3:

– Students discuss and evaluate each other's approaches and ideas.
– Students build and evaluate models of real-world situations.
– Students look for patterns, making and testing conjectures.

Generating possibilities

The issues highlighted by the challenges and opportunities listed above can be summarized as follows: what kinds of things can we do to make the task more meaningful, open it up to discussion so that students engage in collective sense-making rather than implementing a formula, and provide opportunities for connecting ideas?

Rather than work through these issues hypothetically, we can point to a case example. A teacher we've called Ms. Sierra wanted to achieve precisely the kinds of opportunities described in the previous paragraph. She created a "contextual problem" in which students were supposed to use what they knew to build a model of what happened in a realistic scenario – one in which students could use their real-world knowledge while pursuing meaningful mathematics. Here is the task:

> Anay buys a car for $5,000. The car loses 15% of its value every year.
>
> a.   How much is the car worth after 1 year?
> b.   Write an equation to model the value of the car over time.
>
> *Before you go on, find a way to check/justify that your equation is realistic and show your work.*
>
> c.   How long before the car is worth half of its original value?
> d.   After owning the car for ten years, it breaks down. Anay finds out that she will need to replace the clutch to be able to drive the car again. Is it worth it?

Ms. Sierra gave this problem as an in-class assignment. The class worked on it in small groups and Ms. Sierra convened the class as a whole when it seemed profitable or useful to do so (e.g., when many groups were stuck at the same point, a whole-class discussion pointed to a profitable direction without giving away too much information). The extended classroom discussion of the problem did indeed address many of the opportunities identified above.

The discussion of the "car value problem" in Ms. Sierra's class touched on many issues related to all five TRU dimensions, not just Dimension 1. For an extended discussion of the lesson, see *Helping Students Become Powerful Mathematics Thinkers: Case Studies of Teaching for Robust Understanding* (Schoenfeld et al., 2023).

Postscript

We've provided these examples to illustrate two things. The first is how we use all three targets from one dimension when trying to enrich tasks from the curriculum. Looking at the actual targets, and thinking about limitations of current tasks and how they might be opened up, is essential. In practice, once you've done this a few times, you can identify relevant opportunities quickly.

Second, we provided examples at the elementary, middle, and secondary levels to show that the targets are helpful at all grade levels. We'll continue doing that for the remaining four dimensions of TRU.

Notes

1   We had fun creating the example above. All four graphs are identical.
2   There are many other possible goals, such as making connections between graphs and real-world phenomena. For reasons of space, we're only pursuing one possible direction.

4

Dimension 2
Cognitive Demand

What the cognitive demand dimension involves

In order to grow mathematically, students need to make sense of new ideas and integrate them into what they know.

Little meaningful learning takes place when students are deprived of opportunities to engage in mathematical thinking and sense-making. If students are consistently spoon-fed mathematics in bite-sized pieces or told how to solve problems whenever they run into difficulties, they're being deprived of opportunities to build deep understandings and productive habits of mind. The same is the case when students are asked to work on problems that they have not had adequate opportunity to understand or that are so far beyond what they know that they are not in a position to engage productively with the tasks. The goal is to use tasks and create classroom environments that support meaningful engagement and "productive struggle." That's the issue of cognitive demand.

The right level of scaffolding helps students to confront the challenges they encounter, leaving them room to make their own progress on those challenges. What makes things difficult (and interesting!) is that every student in every class has different strengths and needs. The challenge is to find ways to "meet students where they are," leverage individual students' strengths, and help move all students forward toward intended learning outcomes.

DOI: 10.4324/9781003376903-4

A bit more background

Teaching so that students can develop proficiency in the mathematical practices – problem-solving, reasoning, modeling, making mathematical connections, etc. – calls for giving students room to engage with complex mathematical ideas. This, in turn, calls for a broader range of classroom activities than found in traditional texts.

For many decades, mainstream teaching in the US consisted of what Lappan and Phillips (2009) call "show and practice" instruction – the teacher demonstrates and explains a particular procedure, after which students practice working on a collection of similar examples. Traditional textbooks in the US were designed to support this approach. Student editions of mathematics textbooks contained "two-page spreads" of exercises, and the teacher editions of those texts surrounded the exercises with worked examples for the teacher and suggested homework assignments. In the traditional system, expectations were straightforward. The "show and practice" model dominated from the 1950s through at least the 1990s.

The move toward a broader range of classroom activities and goals catalyzed by the NCTM standards documents (NCTM, 1989, 2000) and the Common Core (CCSSI, 2010) placed increasing demands on both students and teachers. When students don't know what to do with challenging tasks, it's hard on them, and it's hard for teachers to see students struggle unsuccessfully. A natural response documented by Stein, Henningsen, and colleagues (Henningsen & Stein, 1997; Stein, Engle, Smith, & Hughes, 2008; Stein, Grover, & Henningsen, 1996) is for teachers to scaffold away some of the difficulties – e.g., "Why don't you try this?" But if that scaffolding removes most of the challenges, then students don't have the opportunity to engage in productive struggle. They don't build their own understandings, they don't learn to persevere, and they don't learn that they can solve complex problems with persistent effort![1]

This issue is difficult for students and teachers. Students develop their understandings regarding the nature of mathematics and expectations for how they are to engage with it from their classroom experiences. If those experiences are deep and engaging, students develop robust understandings of mathematics and positive mathematical identities. But if those experiences are not, the lessons students learn can be problematic. Summarizing the literature on some of the unintended consequences of traditional instruction, Schoenfeld (1992, p. 359) listed the following "Typical student beliefs about the nature of mathematics":

– Mathematics problems have one and only one right answer.
– There is only one correct way to solve any mathematics problem – usually the rule [method] the teacher has most recently demonstrated to the class.

- Most students cannot expect to understand mathematics; they expect simply to memorize it and apply what they have learned mechanically and without understanding.
- Students who have understood the mathematics they have studied will be able to solve any assigned problem in five minutes or less.

Many of these beliefs, developed as a result of classroom experience, are common today. They will only change when students have managed to engage, repeatedly and successfully, with complex problems. Working toward the "sweet spot" of cognitive demand is a key component of helping students learn to engage in productive struggle.

Some ideas that may help

The goal is to arrange things so that students are being stretched, supported if necessary, and able to put things together for themselves. There are at least three ways to work toward this.

The first is to provide students with opportunities to engage with mathematics that is richer, deeper, or more connected. For an example, let's return to the train problem discussed in Dimension 1:

Train A leaves a train station at noon and travels at a steady speed of 50 miles per hour. Three hours later Train B leaves the station on a parallel track, traveling at a steady speed of 60 miles per hour. How long does it take for Train B catch up with Train A?

When given as a standard algebra exercise, this task has an "either you can do it or you can't" quality, depending on whether the student knows how to translate the situation described in the problem statement into algebraic form. Some students will be able to make their way into the problem, and by working on it they will refine their understandings. That's what we hope for. But some students will be lost and some will be bored.

The modified version of the problem, in which students are told to solve the problem using whatever method(s) they find comfortable and in which the class compares and contrasts methods, allows students to pick their preferred methods of solution. Some students make tables, some draw graphs, and some use equations. This opens the problem up, making it more accessible. More students, potentially, are able to engage with it. But it's the follow-up questions teachers can ask that help stretch students and increase the cognitive demand. For example:

- Can you see Train B catching up with Train A in the tabular solution? In the graphical solution? In the algebraic solution?
- If you are looking at a table, can you find a pattern that helps you predict when Train B will catch up with Train A? Can you do the same by looking at a graph? Can you see Train B catching up to Train A in the algebraic solution to the problem?
- Different tables (or graphs) the class has produced have Train A start at $t = 0$ hours or $t = -3$ hours. What is the advantage of doing either one? Why?
- How do the algebraic approaches reflect a starting time of $t = 0$ hours or $t = -3$ hours?

The students who begin their work with tables can see patterns and learn more about how those patterns are visible in the other representations. The students who treat the original problem as an algebraic exercise may well find themselves challenged to describe where "catching up" shows up in their solution, or why there is an advantage to defining t so that you are solving the equation

$$50(t + 3) = 60t$$

instead of the equation

$$50t = 60(t - 3)$$

In this way, deepening the mathematics that students engage with provides additional opportunities for sense-making. The students start working with ideas that are accessible to them, but then they are stretched. This version of the problem provides more opportunities for productive struggle.

A second way to adjust cognitive demand is by scaffolding – carefully. If a student is stuck or makes a mistake, we're often tempted to suggest a "better" way to approach the problem or explain where the student went wrong. But that may mean we've done the sense-making for the student, removing cognitive demand in the process. The question is, can we render the problem within reach without giving away too much? Often a suggestion like "can you draw me a picture of that?" or "can you tell me what's happening here?" will open things up, as will questions about using different representations. If a student's work appears to be based on a misconception, asking "why is that true?" may generate an explanation and a self-correction; asking the student to consider an example that contradicts what they have written or said may open things up. For a large number of examples, take a look at the Formative Assessment Lessons (FALs) at http://map.mathshell.org/lessons.php. Each lesson has a table of "common issues" – difficulties that students often encounter or mistakes they often make – and "suggested questions and prompts," ways of getting students to engage more deeply without telling them what to do next.

Third, a mix of classroom activities can provide extra opportunities for students to engage with the content. For example, activities such as "think/ pair/share" can support emergent bilingual students to talk with each other in their preferred language and allow students time to get their heads around the content. Having students make posters and either do a "gallery walk" or exchange ideas with students at neighboring tables can anchor the mathematics in the students' own thinking. These methods also take advantage of the fact that students can help each other. This eases the burden of trying to address every student's questions.

More specific questions for planning and review

Here are some specific questions you might think about as you consider the opportunities for productive struggle provided by your text or lesson plan or reflect on how a lesson went. Some are drawn from the *TRU Math Conversation Guide* (Baldinger, Louie, and the Algebra Teaching Study and Mathematics Assessment Project, 2018).

–   What opportunities exist for students to grapple with important mathematical ideas? How can we provide more opportunities for meaningful engagement?
–   Which kinds of classroom activities seem to provide students room for productive struggle and support once they are engaged?
–   Are students comfortable sharing partial work or tentative ideas as part of a larger conversation? How can we build norms that support students' entering into productive discourse?
–   Are students taking risks in the service of learning, and do they appear comfortable doing so? How can we encourage more risk-taking, helping them persevere when they get stuck?
–   What resources (other students, the teacher, notes, texts, technology, manipulatives, various representations, etc.) are available for students to use when they encounter struggles? Are there more resources we can make available? How might students be supported to make better use of resources?

- Who does the heavy lifting (students or the teacher) of deciding whether a problem has been solved and whether the mathematics in the solution makes sense? How can students be supported to check their own answers and sense-making?

Some resources that might be useful

Appendix A contains a collection of classroom strategies that serve multiple purposes. Specific strategies that can be used to modify the level of cognitive demand include: *What's the big picture?; Always, Sometimes, Never; Ask for patterns and outliers; Invent an argument; Make categories; Open it up; Share wrong answers like they're right answers; Three things; Time alone;* and *Three Reads.*

The SERP website "the deck behind the 5x8 card," at http://math.serpmedia.org/5x8card/deck/, provides links to a number of useful techniques – similarly, for multiple purposes.

NCTM offers a wide range of resources at http://www.nctm.org/. NCTM's statewide affiliates have annual meetings that are informative; other groups such as teachers' math circles explore rich mathematical tasks. The following books and papers might be of interest.

Books that address these issues directly include Horn (2012), Humphreys and Parker (2015), Stein and Smith (2011). The TRU *Conversation Guide* (Baldinger, E. Louie, N., and the Algebra Teaching Study and Mathematics Assessment Project, 2018) and the TRU *Observation Guide* (Schoenfeld and the Teaching for Robust Understanding Project, 2018), available at https://truframework.org/tools/, provide suggestions for enhancing classroom activities along all five TRU dimensions. Articles that flesh out these ideas include Henningsen and Stein (1997); Lappan and Phillips (2009); Schoenfeld (1992); Stein, Engle, Smith, and Hughes (2008); Stein, Grover, and Henningsen (1996).

The cognitive demand targets

Figure 4.1 lists the core questions for the targets; Figures 4.2–4.4 elaborate on those core questions. You may want to articulate your own goals or ideas on the target in Figure 4.5. As with the mathematics targets, look at the tasks you're planning to (or did) use. See if you can place them on the targets and if you can modify them to move more toward the center.

| Cognitive Demand | *To what extent are students supported in grappling with and making sense of mathematical concepts?* |

Core Questions:

1. Grappling with the mathematics. What opportunities do students have to grapple with and make their own sense of mathematical ideas in this lesson? Which students have these opportunities?

2. Challenges and productive struggle. What challenges do students experience with the tasks and activities? What happens when students experience challenges? How does struggle with mathematical ideas support their participation and understanding?

3. Supporting participation. In what ways does the environment support active participation and sense making?

Figure 4.1 Core questions about Dimension 2, cognitive demand.

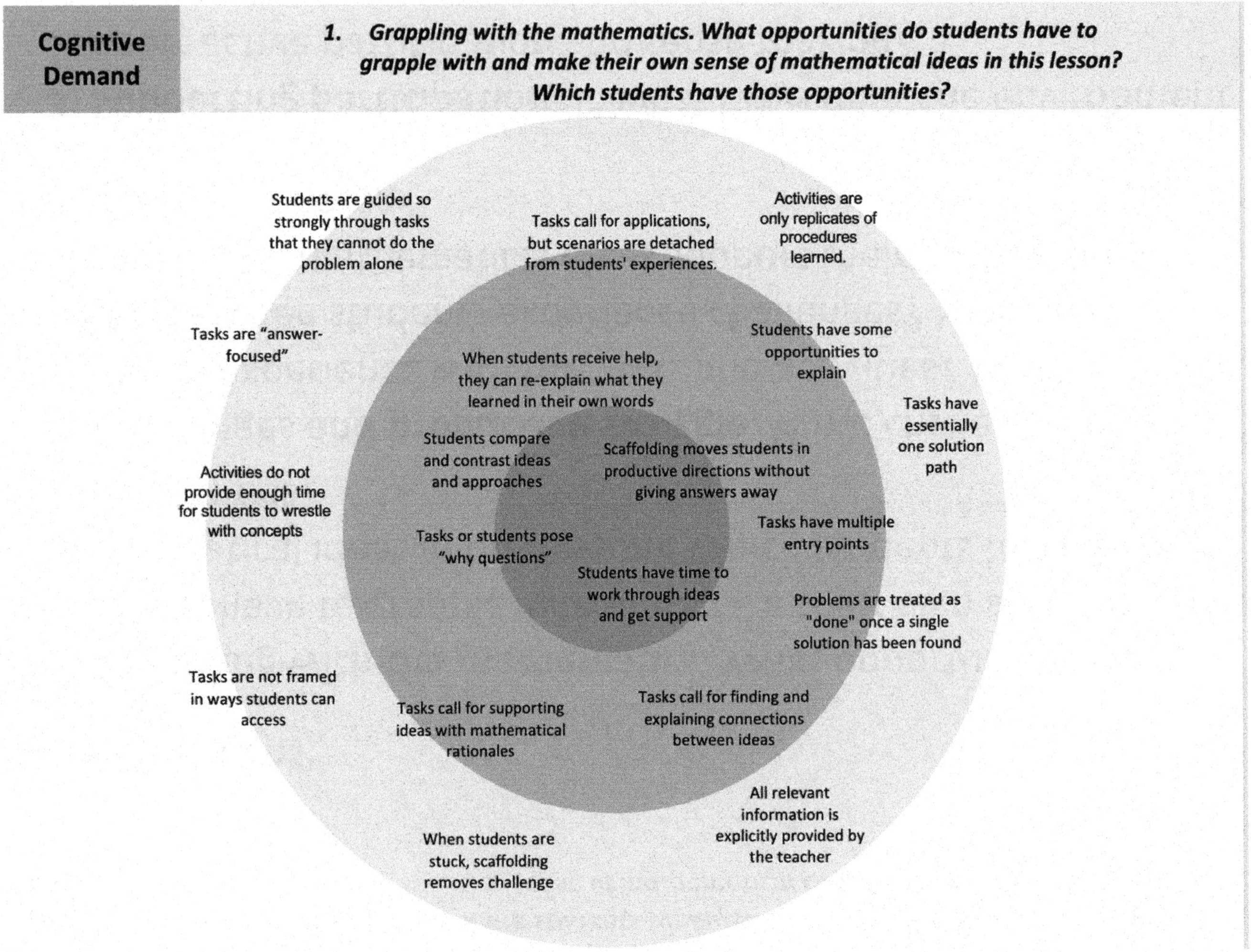

Cognitive Demand

1. **Grappling with the mathematics. What opportunities do students have to grapple with and make their own sense of mathematical ideas in this lesson? Which students have those opportunities?**

Activities are only replicates of procedures learned.

Tasks have essentially one solution path

Students have some opportunities to explain

Tasks call for applications, but scenarios are detached from students' experiences.

Problems are treated as "done" once a single solution has been found

Tasks have multiple entry points

When students receive help, they can re-explain what they learned in their own words

Scaffolding moves students in productive directions without giving answers away

Students have time to work through ideas and get support

All relevant information is explicitly provided by the teacher

Students compare and contrast ideas and approaches

Tasks or students pose "why questions"

Tasks call for finding and explaining connections between ideas

Students are guided so strongly through tasks that they cannot do the problem alone

Tasks are "answer-focused"

Activities do not provide enough time for students to wrestle with concepts

Tasks call for supporting ideas with mathematical rationales

When students are stuck, scaffolding removes challenge

Tasks are not framed in ways students can access

Figure 4.2 Cognitive demand target 1, opportunities for sense-making.

Cognitive Demand

2. Challenges and productive struggle. What challenges do students experience with the tasks and activities? What happens when students experience challenges? How does struggle with mathematical ideas support their participation and understanding?

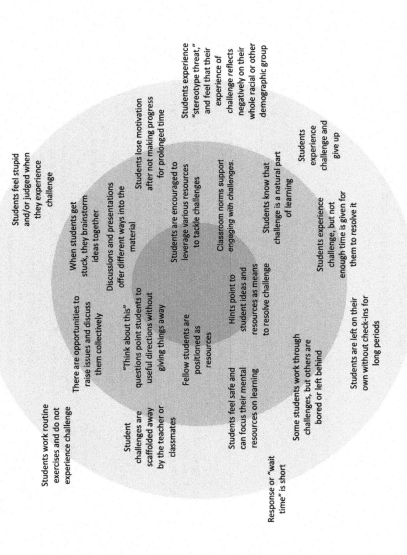

Figure 4.3 Cognitive demand target 2, challenges and productive struggle.

Cognitive
Demand

3. Supporting participation. In what ways does the environment support participation and sense making?

Tasks or discussions are focused on answer-getting and not on solution process

Tasks are focused on procedures and not on conceptual understanding

Some students are "helped" in ways that make status problems worse - they are always expected to need "help"

Students have opportunities to reveal their partial understandings

Incorrect answers are ignored or other students are asked to respond

There is one "right" way to do most problems

Discussions and presentations offer different ways into the material

Students have various ways to share their thinking (speaking, writing, demonstrating, etc)

There are "check-ins" to see if ideas make sense

Perseverance and mindset are discussed as important parts of doing math

Classroom norms support grappling with ideas

Students are positioned as resources for each other

Classroom norms allow for safety in making mistakes

There is time to work through ideas until students see them fitting together

Students raise questions and share interests, but activities do not pursue them in depth.

There is little room to hear out students' thoughts

A handful of students do most of the intellectual heavy lifting

Tasks center on key conceptual issues

There are opportunities to raise issues and discuss them collectively

Attempts at sense-making, for example reasonable conjectures) are valued

Some activities involve exploration with technology, manipulatives, etc., but those activities are not closely tied to core content

Students who need help in one area, such as a particular skill or academic language, are assumed to need help in all areas

Students are left on their own for long periods without check-ins

Authority resides with the teacher or text; student sense making is not expected

Figure 4.4 Cognitive demand target 3, supporting participation.

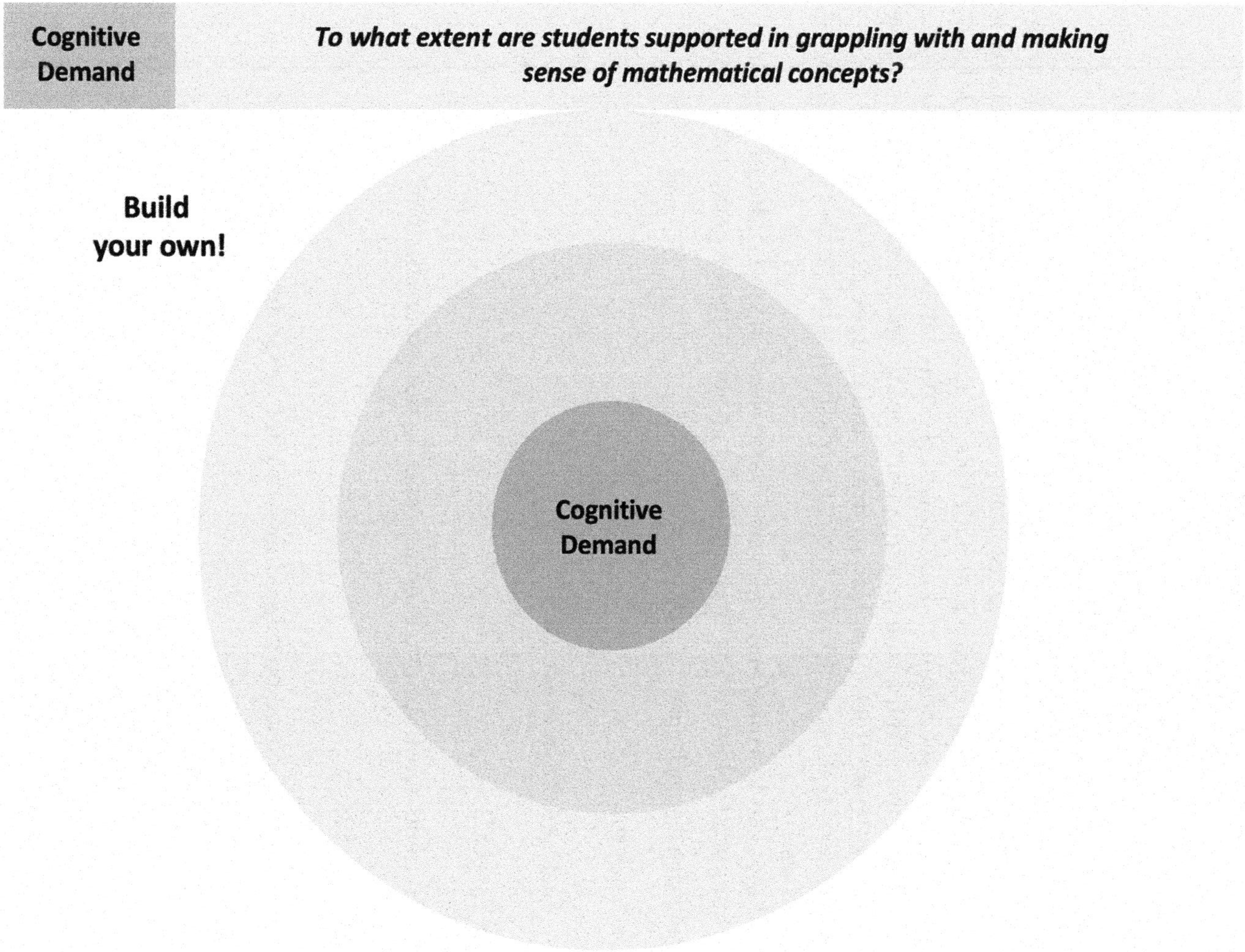

Cognitive
Demand

To what extent are students supported in grappling with and making sense of mathematical concepts?

Build
your own!

Cognitive
Demand

Figure 4.5 Make your own cognitive demand target!

Using the cognitive demand targets – three examples

Here are our cognitive demand examples, in ascending grade order. Each example features one strategy.

An elementary grades example – the border problem

Task Enhancement Strategy: Provide concrete examples as a basis for sense-making

Students need to develop fluency using order of operations to evaluate arithmetic expressions. This topic, typically taught in late elementary school, is difficult to teach in a way that isn't rote or memorization-heavy. Order of operations is a set of rules – mathematical "conventions." Students are typically asked to learn the rules and then practice them. The challenge is, can we help make these rules seem less arbitrary and more reasonable?

Figure 4.6 shows part of a typical worksheet with "order of operations" exercises.

How can learning and practicing basic ideas behind order of operations be enhanced? Here are the challenges and opportunities we see on the targets.

Identifying the challenges

On Target 1:

– Tasks are answer-focused.

On Target 2:

– Students work routine exercises and do not experience challenge.

On Target 3:

– Tasks are focused on procedures and not conceptual understanding.

1) $(11 + 9) \times 8 - 3$	6) $(12 + 7) \times 4 - 9$
2) $7 \times 13 + (8 + 4)$	7) $14 \times 4 + (6 - 3)$
3) $(6 + 11) \times (17 - 9)$	8) $(15 + 4) \times (16 - 6)$
4) $(17 - 6) + 14 \times 4$	9) $(12 - 7) + 11 \times 4$
5) $(16 + 41 - 6) \times 12$	10) $(17 + 21 - 9) \times 4$

Figure 4.6 An "order of operations" worksheet.

Identifying the opportunities

We also identify the following opportunities.

On Target 1:

– Tasks or students pose "why" questions.

On Target 2:

– Students are encouraged to leverage various resources to tackle challenges.

On Target 3:

– There is time to work through ideas until students see them fitting together.

Generating possibilities

It's important for students to learn to make connections between "real world" objects and the symbols that represent them. That way they make concrete connections, rather than simply performing arithmetic or symbolic manipulations. Here, introducing area models provides that kind of opportunity. See the Border Problem in Figure 4.7.

The class can then compare and contrast their solutions. Figure 4.8 illustrates one solution, where the area can be seen as (4 × 4) + 4.

The area of every small square in this figure is 1. Write as many arithmetic expressions as you can that show the area of the border. Illustrate your expressions on the figure.

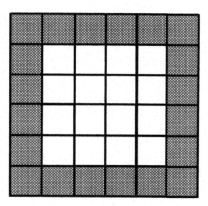

Figure 4.7 The Border Problem.

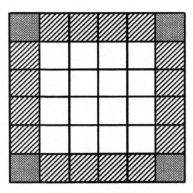

Figure 4.8 Area = (4 × 4) + 4 = 20.

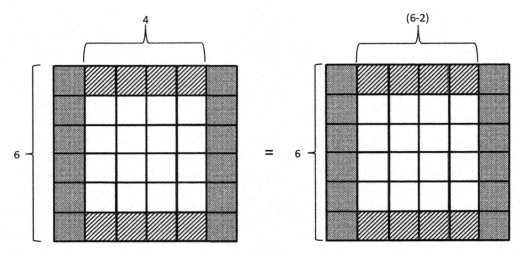

Figure 4.9 Area = (2 × 6) + (2 × 4) = 20 *and* Area = (2 × 6) + 2 × (6 − 2) = 20.

Figure 4.9 demonstrates that a single pattern can be written in two different ways.

There are many more patterns. The students might find 4 × 5, (4 × 6) − 4, or (6 × 6) − (4 × 4), for example. These computations are meaningful because the students can see the numbers in the representations. Expressions that use order of operations incorrectly will not arrive at the correct area, something that students can check arithmetically and visually.

After students generate many ways of finding the area of the border, the teacher holds a whole-class discussion in which students' expressions are elicited, recorded on the board for all students to see, and annotated with how the calculation represents the area of the border. This whole-class discussion supports reasoning about order of operations by connecting the calculations to the diagram.

A middle grades example – the road race problem

Task enhancement strategy: revisiting a lesson to leverage student mistakes

The Mathematics Assessment Project's (MAP) FAL "Representing: Road Race" introduces the idea of slope in middle school by inviting students to make sense of rates of change using representations including story problems, tables, and graphs. In the Road Race activity, students predict where and when two runners running a race on a track at different speeds will lap each other. The activity prompts students to first work in groups and then provides examples of imaginary students' work that uses tables and line graphs to identify where and when the runners will lap each other. See https://www.map.mathshell.org/lessons.php?unit=7200&collection=8.

When pressed for time, teachers and curricula often shorten or eliminate portions of meatier tasks so that the lessons can be squeezed into short class periods or already-full curriculum units. Here is one teacher's experience. To fit the lesson into her scope and sequence, the teacher gave her middle school students just the group-work portion of the activity, shown in Figure 4.10.

The teacher did not plan to have her students make posters of their work, look at or comment on each other's work, or look at the student work samples provided in the lesson, as the MAP activity suggested. This was her first time teaching this activity, and she wanted to devote no more than 30 minutes to it.

A Race

Amy and Rebecca are running in a road race.

The map, drawn to scale, shows the route of the race:

The race consists of four laps of the route and Amy and Rebecca run clockwise along the route at a constant speed.

It takes Amy 8 minutes to run a mile. Rebecca takes 12 minutes to run a mile.

1. Mark on the map where Rebecca 'R' and Amy 'A' will be one hour into the race. Explain how you know.

2. Will one runner 'lap' the other runner at some point in the race? If so, where? Label this place 'X' and explain your reasoning. If not, explain how you know.

Figure 4.10 A task from the Formative Assessment Lesson "Representing: Road Race".

As the lesson unfolded, the students got stuck. The teacher was pulled from group to group putting out fires, leaving her little time to attend to student thinking. She told a few groups of students to make a table or try graphing. Those students followed her instructions but made mistakes on their tables and graphs and gave up. Feeling overwhelmed by the amount of struggle she was encountering and the short amount of time she had allotted, the teacher ended the activity early. She collected students' work, told them that she was proud of how hard they'd worked on a task that she also found surprisingly challenging, and indicated that they'd move onto a new task the next day.

However, the teacher's decision to move on didn't sit right with her. She had hoped that the task would lead to a productive discussion. Also, she felt that abandoning a challenge sent the wrong message to her students.

The following items on the targets identify the challenges as they played out and opportunities for revisiting the task.

Identifying the challenges

On Target 1:

–   Activities do not provide enough time for students to wrestle with concepts.

On Target 2:

–   Students experience challenge and give up.

On Target 3:

–   There is little room to hear out students' thoughts.

Identifying the opportunities

On Target 1:

–   Students have time to work through ideas and get support.

On Target 2:

–   Hints point to student ideas and resources as means to resolve challenges.

On Target 3:

–   There are opportunities to raise issues and discuss them collectively.

Generating possibilities

The kind of difficulty this teacher encountered – students experiencing so much challenge that it's a struggle to find ways to support them – happens even to the most experienced teachers when teaching a new, ambitious lesson. Sometimes it makes good sense to thank students for their hard work and shift to something different. Learning is difficult when the resources provided by activities and classroom routines don't support students to productively work through challenges.

But even when students are "stuck" they may well be generating ideas that are potentially productive. It may be impossible in the heat of moment for a teacher to figure out how to best use those sparks of productive thinking, as it was in this case. Before the next day's class the teacher had time to look over the tables and graphs that some of her students had started making. Inspired by a part of the MAP activity that she hadn't planned to use, in which students looked at imaginary students' work, she decided to distribute copies of several students' partially incorrect but potentially generative work to all of the students during the next class. She hypothesized that if other students saw these representations, they could make progress by trying to correct and continue them.

The next day, she handed out three different tables and graphs that some of the students had begun. She prompted her students to discuss and write about:

1   How they thought each table and graph might help them finish the task
2   What questions they had about each table and graph
3   How they might correct and finish each table and graph

As the students worked on these representations, they began to understand the task better. Trying to fix mistakes they found on the representations led them to ask sense-making questions about the task, which helped them to continue the representations and use them to make progress on the task. To close out the activity, the teacher had students present about what they learned about the task from looking at each other's work. The students shared how the different representations helped them to see the race situation in different ways. This in-depth work supported them answering the original problem.

A secondary grades example – diagonal areas

Task enhancement strategy: use student work to enrich a discussion

Justification and proof are essential mathematical practices, which are often emphasized in high school geometry. Figure 4.11 shows a typical problem, and Figure 4.12 offers one possible solution.[2]

Diagonal Areas

Prove that if you draw in the two diagonals of a parallelogram, you divide the parallelogram into four equal areas.

Figure 4.11 A proof problem.

Diagonal Areas - Solution

In the parallelogram below all the white triangles are congruent and all the gray triangles are congruent.

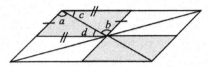

This congruency can be proved using Side, Side, Side (SSS): The diagonals of the larger parallelogram bisect each other. This means the small gray quadrilateral is a parallelogram. Opposite sides of the gray parallelogram are equal. (Note that a similar proof is possible with Angle, Side, Angle).

It follows that the four triangles consisting of one gray and one white triangle are all equal in area. Hence the statement is true.

Figure 4.12 A solution to the proof problem.

Typical solutions involve drawing additional construction lines to create congruent triangles, making an argument that the new, small triangles are congruent, and then using this fact to establish that the larger triangles of interest are congruent. Figure 4.12 provides a condensed version of a proof.

Many students find this task quite challenging! Although different proofs of congruence are possible, the problem has essentially one solution path, which involves seeing that the four parts of the parallelogram created by the two diagonals each consist of one white and one gray triangle. Giving students hints is, in essence, nudging students along that one path. Let's look at the targets to see if the task can be broadened to include other aspects of mathematical thinking.

Identifying the challenges

On Target 1:

–   Tasks have essentially one solution path.

On Target 2:

– Students experience challenge and give up.
– Student challenges are scaffolded away by the teacher or classmates.

On Target 3:

– Authority resides with the teacher or text; student sense-making is not expected.

Identifying the opportunities

On Target 1:

– Tasks have multiple entry points.

On Target 2:

– Discussions and presentations offer different ways into the material.

On Target 3:

– Students are positioned as resources for each other.
– Classroom norms allow for safety in making mistakes.

Generating possibilities

Two changes would enhance the task significantly to realize the possibilities suggested by the targets. Both of these changes are exemplified in the FAL, "Evaluating statements about length and area." The first change is to expand the task. Instead of giving a known statement about a particular type of quadrilateral, the FAL invites students to explore and justify their reasoning about various types of quadrilaterals. See Figure 4.13.

Second, we can ask students to critique work done by other students. In the "length and area" FAL, after considering the enhanced diagonal areas

---

Diagonal Areas (Enhanced Version)

Is the following statement sometimes, always or never true?

"If you draw in the two diagonals of a quadrilateral, you divide the quadrilateral into four equal areas."

Figure 4.13 An enhanced version of the diagonal areas task.

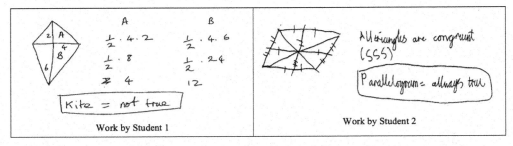

Figure 4.14 Work to be improved. Doing such tasks reveals misunderstandings and provokes rich conversations.

problem in small groups and as a whole class, students are given samples of work done by other students and asked to help the author "improve" them (See Figure 4.14). The students are asked,

- What do you like about the work?
- Has this student made any incorrect assumptions?
- Is the work accurate?
- How can the explanation be improved?

Working tasks like this helps students develop critiquing and reasoning skills, which are important for self-critiques and for advancing classroom dialogue.

Attention to classroom norms is critical here. In addition to the learning opportunities for mathematical practices, activities like this can *potentially* help create productive norms around the value of incorrect and partial answers, safety in making mistakes, and respectful critique and revision of each other's work. However, as with any activities that ask students to take increased intellectual risks, it is essential to pay attention to dynamics of participation, inclusion, and exclusion as they unfold in the classroom.

Notes

1   On the flip side, there's also a positive feedback loop when students are supported in productive struggle. Teachers and students are often happily surprised by what students can do when they're given some room to try out ideas. Once they discover this, they are increasingly willing to take risks. See Research for Action (2015).

2   The ideas in this example are drawn from the Formative Assessment Lesson "Evaluating statements about length and area," which can be downloaded at https://www.map.mathshell.org/lessons.php?unit=9310&collection=8.

5

Dimension 3
Equitable Access

What the equitable access dimension involves

Who participates in classroom activities and in what ways? Do all students have opportunities to engage with important content and contribute to collective discussions? How is diversity valued, and how are students' widely divergent backgrounds and experiences leveraged to create rich learning opportunities for all students?

We ask these questions when we think about the extent to which our classrooms provide students equitable access. In classrooms that attend to equitable access, more students have more opportunities to engage with central and challenging mathematical content. Moreover, teaching attends to students' diverse needs and provides students with scaffolds, resources, and participation structures that build on their strengths and support their engagement with that content. Eliciting contributions from a wide range of students offers opportunities for a broader range of ideas to be discussed, contributing to mathematical richness.

There is a long history of differential achievement ("performance gaps") by students who come from varied racial, ethnic, and economic backgrounds. And there is strong evidence that such differences are tied to "opportunity gaps" – differential access to opportunities to learn (Oakes, Joseph, & Muir, 2004). Opportunity gaps occur both at a "macro" level, for example, due to inequitable distribution of resources across a state or district or due to tracking

DOI: 10.4324/9781003376903-5

or teacher assignments within a school, and at a "micro" level, in classrooms. While issues of access between classrooms are critically important (see, e.g., Schoenfeld, 2022), they are beyond the scope of what we can address in this section, which focuses on the actions individual teachers can take within their classrooms to provide equitable access to mathematics. Do all students have frequent opportunities to discuss important ideas? In *How Schools Shortchange Girls* (AAUW, 1992), for example, research revealed that in math classrooms boys were being called upon far more often than girls. Moreover, when girls were called upon, they were often asked questions that were less conceptually oriented than the questions that boys were asked. That kind of pattern denied girls meaningful opportunities to engage in key mathematical practices. Similar inequities take place when students are separated from others to do remedial work while the rest of the class advances – the students who are supposedly "catching up" are denied opportunities to progress with the rest of the class. Such inequities occur with regard to gender, ethnicity, and race – see, e.g., Reinholz and Shah (2018).

We stress that participation is a social activity – and that classrooms are social environments that, in many ways, tend to reflect the larger social and cultural environment. Absent teacher interventions, typical patterns of participation, inclusion, and exclusion from the wider society tend to "push into" classrooms. To give one example, there is much more to the question of equitable access than giving students an equal number of conversational "turns." What matters is who gets to do what, how students are positioned by others, and even how students are listened to and responded to. Participation patterns over time, in both whole class and small group settings, are important. Do specific people or groups get more "air time" on central mathematics than others? If so, there's an issue.

Within classrooms, equitable access to mathematics means that every student has the opportunity to engage in meaningful ways with the main mathematical content and practices of every topic over the course of the year. For classroom activities to be equitable, all students must have opportunities to work with the big ideas and patterns of reasoning, sense-making, and problem-solving that are necessary for them to develop into knowledgeable and flexible mathematical thinkers and problem-solvers. This means that we have to be careful to not privilege particular aspects of mathematical performance over others – for example, privileging speed may disadvantage students who are deeply thoughtful but not especially quick. Or, not taking time to address issues of vocabulary or problem context may leave some students unable to participate verbally. There may be rich discussions or other productive activities taking place. But, if all students do not contribute to and benefit from these activities, our classrooms do not provide equitable access. Also, as we observed in our discussion of Dimension 1, increasingly rich mathematics (e.g., employing multiple representations and different ways of viewing the same mathematical phenomena) can also offer additional access points into mathematical sense-making.

A bit more background

Equitable access does not mean that all students are doing the same things all the time. In fact, "sameness" is not a guarantee of high-quality instruction or equity: if all the students are working low-level worksheets, then none of the students are well served. It is essential for every student to have opportunities to engage with rich and meaningful mathematical content and practices.

Asking the following questions can help us reflect on some issues of access:

Are some students separated from the rest of the class to work on prerequisite mathematics for long periods of time, in a way that excludes them from learning the central mathematical content? Do tasks have a high linguistic threshold, placing some of them beyond the reach of emerging bilinguals? Do some students get called on consistently, while other voices are never heard? Do some students feel they have space to contribute, while others do not?

Similarly, reflecting on the following questions can help us think about ideas that might inspire more equitable access:

Are there multiple ways to interact with the content, so that students with different strengths can participate fully and learn from each other? Does every student have the opportunity to conjecture, to venture ideas, to have their ideas considered, and to comment on others' ideas?

Our goal is to invite and then bring as many students as possible into interaction with powerful mathematics, while at the same time creating opportunities to build on students' strengths and address students' needs.

Do particular students have knowledge or interests inside or outside of school that can be leveraged to enrich the mathematical conversations? Can some aspect of students' lived experience be mathemetized in ways that enrich classroom mathematical conversations?

Equitable classrooms provide each and every student access to meaningful disciplinary concepts and practices, supporting each student in developing their own understandings and building productive disciplinary identities (Dimension 4). There may be rich discussions or other productive activities taking place – but the general question is, who contributes to those discussions or activities, and who profits from them in what ways?

Some ideas that may help

Effective teachers use a range of strategies that create robust classroom discourse communities and encourage meaningful participation by all students (Boaler & Staples, 2008; Cohen & Lotan, 1997, 2014; Schoenfeld, 2003). These teachers establish and reinforce expectations for various forms of participation in and contributions to classroom activities. They utilize tasks that offer opportunities for students to grapple with challenging and interesting content, and they provide support in ways that enable all students to engage in meaningful ways (Horn, 2012; Nasir, Cabana, Shreve, Woodbury, & Louie, 2014).

Some of the ways we can work to enhance equitable access to rich mathematical content for our students include the following.

Establish and maintain classroom norms that create safe spaces for academic risk-taking.

Typically, teachers set classroom rules or norms early in the school year to establish their expectations for student behavior. Some norms, such as *Be Respectful* and *Be Honest*, should exist in all classroom settings. It's important for students to feel safe when they venture ideas in the classroom. Learning environments need to support academic risk-taking so every student has access to meaningful mathematical engagement.

Mathematical sense-making involves venturing new, incomplete, and/or tentative ideas. Making mistakes is part of the process. It's how progress is made – and that's how students should feel when they're working on challenging tasks.

Classroom norms can make a big difference in creating a safe space for all students to feel comfortable taking academic risks as they try out new ideas, work through confusion, and build new understandings. A few examples of "safe space" norms are: *Mistakes are Valuable, Questions are Crucial, Depth is More Important than Speed, We Get Smarter Together.*

We can be explicit about saying which norms we want students to attend to and reinforce those norms – e.g., when a student makes a mistake or asks a question. For example, using sentence starters (e.g., saying "I disagree, because..." instead of "you're wrong") can help to establish productive and respectful norms for group discussions. When students feel comfortable contributing ideas, the mathematical conversations that result are richer, deeper, and more rewarding.

In addition, there are more specific norms that can support mathematical collaboration and sense-making.

Encourage and support mathematical contributions by each and every student.

Attending to participation patterns can tell us a lot. Who contributes to the mathematical work, in which circumstances and in which ways? Whose ideas are taken up and built on? Who helps to organize the group's work? Who has ideas to share but seems reluctant to jump in? How are mathematical contributions regulated and supported? Equitable participation will not necessarily occur naturally, either in whole class conversations or in small groups – there are lots of ways in which students can be positioned as strong or weak or as someone worth listening to or not (see Esmonde, 2009).

Noticing specific status differences or inequitable participation patterns (e.g. observing at the beginning of the year that a few white male students tend to dominate the conversation) and working to counteract these patterns in the short, medium, and long term is essential for supporting all students' engagement with rich mathematics.

Various tools can help to provide opportunities for meaningful mathematical participation. However, using any of these tools does not alone guarantee equitable access to content. The following tools are intended to disrupt unfair and unjust power imbalances among students (often aligned with gender and/or race), but if used without careful attention these tools can end up perpetuating persistent inequities (Fink, 2022). Explicit attention needs to be paid to how students are being positioned with respect to mathematical competency and each other, with the goal being that every student be positioned as a valuable mathematical contributor to the group's collective learning (see point 3 below).

- Posing questions using a *think/pair/share* structure encourages each student to contribute by giving time and space for students to form their own ideas about a particular topic, practice articulating their ideas with a partner, and expand their thinking by hearing other students' ideas.

- *Whip-arounds* encourage verbal participation by allowing every student to share their thinking. With this technique, students are asked to write down a short response to a prompt and then take turns one-by-one reading their responses (quickly) to the whole class. (Here too, it has to feel safe to participate – for example, "I pass, Carla said what I was going to say" can be a perfectly fine answer.)
- *Participation chips* can be used during small group conversations to regulate and distribute individual students' contributions. Each student starts with a small number of chips to "spend" on discussion turns. When they're out of turns, they have to wait until everyone has spent their chips and the chips are re-distributed.
- Poster presentation *gallery walks* can provide opportunities for specific students to take on the role of mathematical experts. During the *gallery walk*, one student from each poster group is designated as the group's "explainer," staying by the group's poster to explain their group's learning and answer questions asked by their peers who are rotating.
- Complex Instruction employs specific techniques such as *multiple ability treatments* and *assigning competence* to address status issues in mathematics classrooms (Cohen & Lotan, 2014; Cohen, Lotan, Scarloss & Arellano, 1999).
- *Sharing what your partner said*: instead of sharing their own ideas, have students talk in pairs before a whole class discussion and then share an idea they heard, rather than their own idea.
- Tools such as EQUIP (Reinholz & Shah, 2018) provide ways to track classroom participation for individual students and groups of students (e.g., by gender and / or race), helping to uncover problematic participation patterns.

Support broad notions of mathematical competence.

There are many ways to be smart mathematically, including: making sure you understand the problem, posing interesting questions, keeping track of useful ideas, revising ideas or approaches in the light of new information, facilitating productive conversations, making interesting connections, working systematically, clarifying ideas and expressing them logically, and being able to summarize or explain your own or your group's ideas

(Cohen & Lotan, 2014; Cohen, Lotan, Scarloss & Arellano, 1999; Horn, 2012; Nasir et al., 2014; Mason, Burton, & Stacey, 2010). A narrow definition of being smart in math (e.g., being fast at calculating answers) privileges a subset of students and means that the talents and skills of other students go unrecognized and unrealized. Supporting broad notions of mathematical competence encourages more students to share their ideas and enriches classroom discourse by broadening the range of student contributions. Publicly acknowledging students for their various mathematical competencies rather than calling attention to their deficits encourages more students to contribute to mathematical discourse and to see each other as valuable resources in the learning process.

Provide opportunities for students to share different perspectives and ways of thinking mathematically.

In addition to expanding notions of mathematical competence, we also want to encourage broader participation by valuing students' various points of view and by providing opportunities for students to share different ways of thinking. Interestingly, *richer* mathematics can often be more accessible to more students! The idea is to open up the space of mathematical thinking so that more students can engage productively. Avoiding a steady diet of known-answer questions (e.g. what is the answer to problem 7, "$10 = 2x + 4$?") and asking more open-ended questions (e.g. "How did you solve problem 7? Did anyone solve it in a different way? Did anyone make a mistake along the way that they think others could benefit from knowing about? How do you know your answer is or isn't reasonable?") can open up the mathematics and make it more interesting. Working problems that draw upon everyday and/ or social justice contexts familiar to students can be motivating and enrich the mathematics (see, e.g., Gutstein & Peterson, 2013). Discussing problems that offer multiple points of access and/or have solutions that can be achieved using multiple representations can open up the mathematics and the possibilities for engagement – recall, for example, the "train problem" discussed in Dimension 1, in which tabular, graphical, and algebraic solutions could all be linked. Implementing "groupworthy" tasks (i.e. tasks that cannot be completed sufficiently by an individual and instead are worth the attention of an entire group) provide chances for all students to engage in the mathematics in rich ways (Cohen & Lotan, 2014).

In geometry, one useful technique is to expand proof tasks. Instead of saying "Prove that X is true" you can say "My friend says X is true. Is it? If it is, provide a proof. If it isn't, give an example showing why." This framing of the task opens it up to investigation and discussion. Even richer are "always, sometimes, never" problems. Here is a simple example.[1]

> James says, "If you draw two shapes, the shape with the greater area will also have the longer perimeter." Is James' statement Always, Sometimes, or Never True? Fully explain and illustrate your answer.

When problems are framed in this way, there is a great deal of room for students to contribute ideas.

Provide support as needed, to remove barriers to participation.

A lack of vocabulary, unfamiliarity with the problem context, or not having particular prerequisite knowledge or skills may diminish the likelihood or quality of students' contributions during class activities. The goal is to remove barriers to participation – not by making the math easier but rather by offering temporary supports that help every student engage fully with core ideas.

As one example, consider the situation that arises when the class is working on complex word problems, and some students are not proficient at the arithmetic operations needed to arrive at solutions. A standard approach is to give the students "pull out time" – to have them practice arithmetic while the rest of the class is working on the more complex problems. But this keeps them away from the kinds of sense-making the rest of the class is doing! One way to deal with this issue is to begin class with a number talk relevant to the particular arithmetic skills needed for the day's work. This helps review and explicitly teach arithmetic strategies, while also reinforcing classroom norms for participation and keeping the learning focus on problem-solving. The goal is to build skills along the way, while not interfering with ongoing learning and development.

Providing scaffolding support for emerging bilinguals, students with IEPs, and everyone else.

The linguistic complexity of word problems or modeling tasks often functions as an obstacle preventing students from understanding the problems well enough to grapple with the core mathematics in those tasks (Moschkovich, 2012, 2013). Moreover, many students fluent in English face similar challenges making sense of problems. Thus, many techniques developed for emerging bilinguals turn out to be useful for all students.

Seemingly straightforward problems can hide linguistic and other traps (see, e.g., Rehmeyer, 2014, pp. 22–27). Consider this problem, taken from a middle school text.

The Java Joint wishes to mix organic Kenyan coffee beans that sell for $7.25 per pound with organic Venezuelan beans that sell for $8.50 per pound in order to form a 50-pound batch of Morning Blend that sells for $8.00 per pound. How many pounds of each type of bean should be used to make the blend?

The linguistic challenges in this and other textbook problems (which referred to "savings bonds," "fungicide," "red pigment," and "processing a 24-exposure roll of film") are tremendous. Moreover, the challenge in these problems goes beyond vocabulary: it goes to understanding the context well enough to be able to represent and/or model it mathematically. All students can be supported by paying attention to the linguistic and contextual challenges in the tasks they are asked to work.

A number of techniques help emerging bilinguals in particular to develop relevant understandings – and these techniques are much more generally applicable. Zwiers et al. (2017) offer both principles for classroom interactions and a series of productive classroom routines to support language development and use. Those routines include the following:

Stronger and Clearer Each Time. In English Language Arts, students do "pre-writes," getting their core ideas on paper, then organizing them for first drafts, and then revising those drafts. We can do the same in mathematics. Each rewrite improves the reasoning and presentation of the argument made in the previous draft.

Collect and display. The teacher can orchestrate conversations, identifying various contributions and "revoicing" them – asking students for clarifications and then making sure the whole class "hears" the ideas so they can be built on. In revoicing, the teacher often gives students credit for their ideas, thus building the students' mathematical identities (see Dimension 4).

Critique, Correct, and Clarify. Consider giving students sample work to critique and improve if necessary. This helps everyone refine their analytic as well as linguistic skills. (It's often easier to spot flaws in an argument someone else produces rather than in one you produce.)

Three Reads. In this strategy small groups of students work through a problem statement three times, with different goals each time. The first time is to make sure they understand the text itself – can they paraphrase the situation, without using numbers? The second time they analyze the language involved and make sure they understand the mathematical structure of what is described in the problem statement. The third time they read the problem statement to brainstorm possible solution methods and discuss their relevance.

Last but definitely not least, discussions can build on the understandings that students bring using their home language (see, e.g., Rosebery, Warren, & Conant, 1992). Providing students with opportunities to discuss ideas in their home language can support essential aspects of sense-making, create a safe space for academic risk-taking, and support broader conceptions of what is valued in the classroom.

More specific questions for planning and review

Here are some specific questions you might consider as you think about supporting more equitable access. These questions are taken from the *TRU Math Conversation Guide* (Baldinger, Louie, and the Algebra Teaching Study and Mathematics Assessment Project, 2018).

- What opportunities exist for each student to participate in the intellectual work of the class?
- What is the range of ways that students can and do participate in the mathematical work of the class (talking, writing, leaning in, listening hard, manipulating symbols, making diagrams, interpreting text, using manipulatives, connecting different ideas, etc.)?
- Which students participate in which ways? Which students are most active, and when?
- In what ways can particular students' strengths or preferences be used to engage them in the mathematical activity of the class?
- What opportunities do various students have to make meaningful mathematical contributions?
- What are the language demands of participating in the mathematical work of this class (e.g., academic vocabulary and mathematical discourse practices)?
- How are norms (or interactions, lesson structures, task structure, particular resources, etc.) facilitating or inhibiting participation for particular students?

- What teacher moves might expand students' access to meaningful participation (such as modeling ways to participate, holding students accountable, and pointing out students' successful participation)?
- How can we support particular students we are concerned about (in relation to learning, issues of safety, participation, etc.)?

Some resources that might be useful

Appendix A contains a collection of classroom strategies that serve multiple purposes. Specific strategies that are often useful for opening up mathematical access in the classroom include *share what your partner said, fishbowl, open it up, wait a turn, what do you wish you knew, share wrong answers like they're right answers, three things, extended wait time,* and *idea melting pot.*

Many of the texts referenced above, e.g., Cohen & Lotan (2014), Gutstein & Peterson (2013), Horn (2012), Mason, Burton & Stacey (2010), Nasir et al. (2014) and Zwiers et al. (2017) contain practical suggestions. Some of our references frame the big issues – access to mathematics as a civil rights issue (Moses, 2001), savage inequalities (Kozol, 1992), and tracking (Oakes, 2005). And there's a lot more. See Boaler & Staples (2008), Burns (2004), Cohen & Lotan (2014), Cohen, Lotan, Scarloss & Arellano (1999), Darling Hammond (2010), Daro (2021), DiME Center (2007), Esmonde (2009), Ladson-Billings (1994), Ladson-Billings & Tate (2006), Lindberg, Hyde, Petersen, & Linn (2010), Martin (2009), Moschkovich (2012, 2013), Moses (2001), Nasir & Cobb (2007), Nasir et al. (2014), National Research Council (2001), Rosebery, Warren, & Conant (1992), Schoenfeld (2003), Shah (2017), Steele (1997), and Steele & Aronson (1995).

The equitable access targets

You know the drill by now! Figure 5.1 lists the core questions for the targets; Figures 5.2–5.4 elaborate on those core questions. You may want to articulate your own goals or ideas on the target in Figure 5.5. Take a look at the tasks you're planning to (or did) use. See if you can place them on the targets and if you can modify them so that they move more toward the center.

If you can do this with a colleague – even better, if the colleague can visit your class and take notes – then you can profit from having a spare set of eyes and comparing notes.

| Equitable Access to Mathematics | *To what extent are all students provided opportunities to engage with the core content and practices of the lesson?* |

Core Questions:

1. In what ways do all students have opportunities to engage with the core mathematical content of the lesson?

2. In what ways are specific student needs addressed during classroom instruction?

3. In what ways are diverse student strengths leveraged during classroom instruction?

Figure 5.1 Core questions about Dimension 3, equitable access.

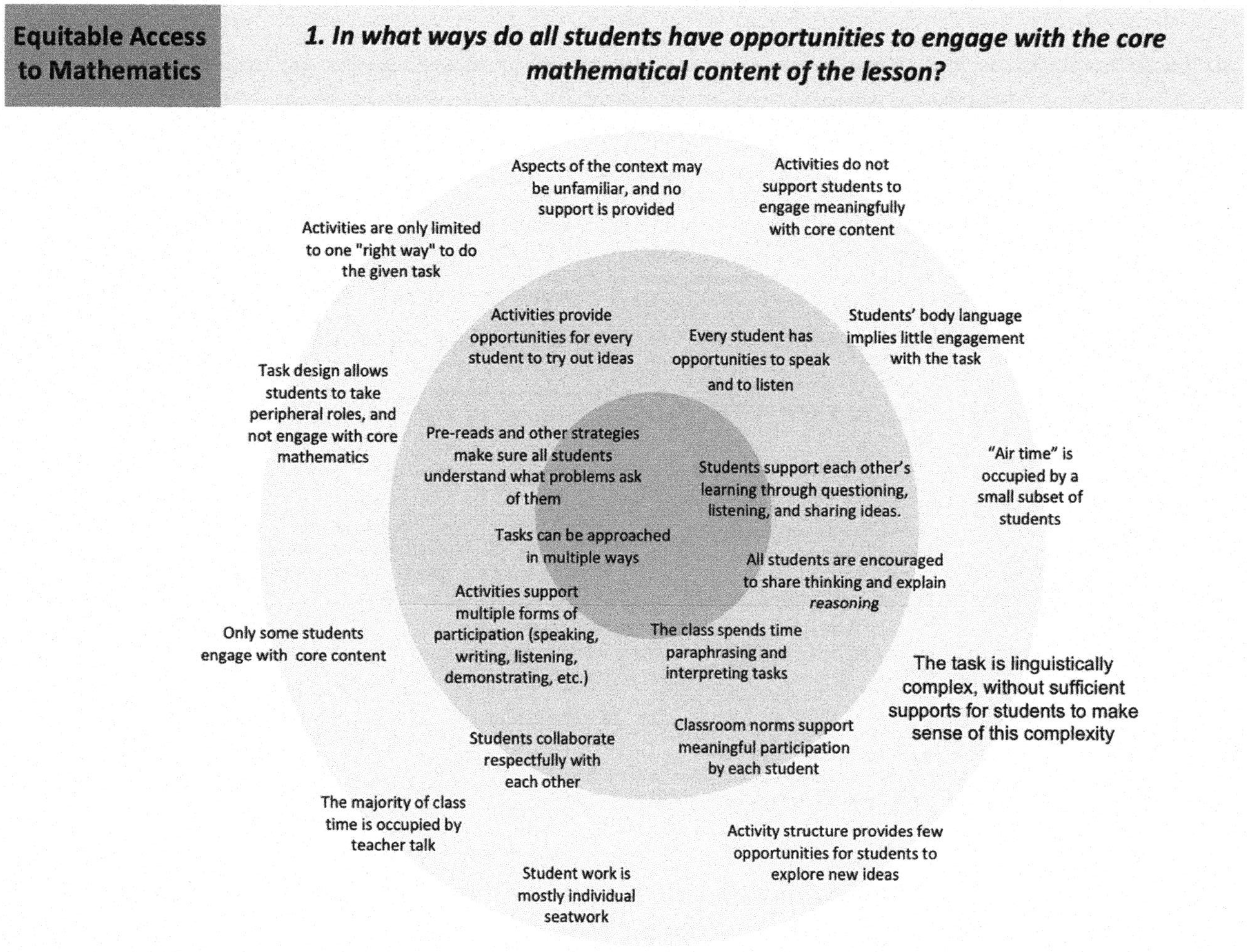

Equitable Access to Mathematics

1. In what ways do all students have opportunities to engage with the core mathematical content of the lesson?

Activities do not support students to engage meaningfully with core content

Aspects of the context may be unfamiliar, and no support is provided

Students' body language implies little engagement with the task

"Air time" is occupied by a small subset of students

Every student has opportunities to speak and to listen

Students support each other's learning through questioning, listening, and sharing ideas.

All students are encouraged to share thinking and explain reasoning

The task is linguistically complex, without sufficient supports for students to make sense of this complexity

Activities are only limited to one "right way" to do the given task

Activities provide opportunities for every student to try out ideas

Pre-reads and other strategies make sure all students understand what problems ask of them

Tasks can be approached in multiple ways

The class spends time paraphrasing and interpreting tasks

Classroom norms support meaningful participation by each student

Activity structure provides few opportunities for students to explore new ideas

Task design allows students to take peripheral roles, and not engage with core mathematics

Activities support multiple forms of participation (speaking, writing, listening, demonstrating, etc.)

Students collaborate respectfully with each other

Student work is mostly individual seatwork

Only some students engage with core content

The majority of class time is occupied by teacher talk

Figure 5.2 Equitable Access Target 1, opportunities to engage with core mathematics.

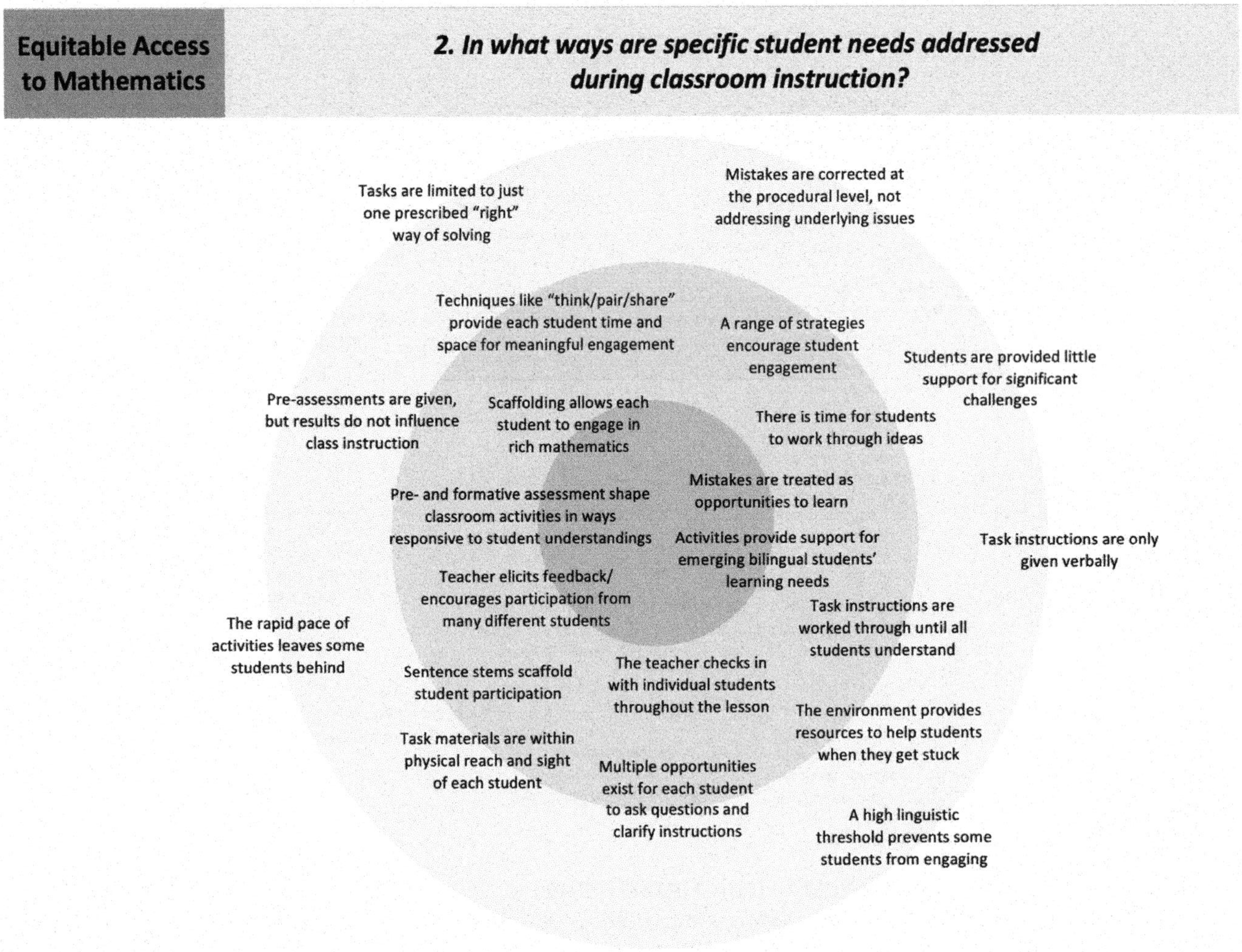

Equitable Access to Mathematics

2. In what ways are specific student needs addressed during classroom instruction?

Mistakes are corrected at the procedural level, not addressing underlying issues

Students are provided little support for significant challenges

Task instructions are only given verbally

A range of strategies encourage student engagement

There is time for students to work through ideas

Task instructions are worked through until all students understand

The environment provides resources to help students when they get stuck

Mistakes are treated as opportunities to learn

Activities provide support for emerging bilingual students' learning needs

A high linguistic threshold prevents some students from engaging

Tasks are limited to just one prescribed "right" way of solving

Techniques like "think/pair/share" provide each student time and space for meaningful engagement

Scaffolding allows each student to engage in rich mathematics

Pre- and formative assessment shape classroom activities in ways responsive to student understandings

The teacher checks in with individual students throughout the lesson

Multiple opportunities exist for each student to ask questions and clarify instructions

Pre-assessments are given, but results do not influence class instruction

Teacher elicits feedback/ encourages participation from many different students

Sentence stems scaffold student participation

Task materials are within physical reach and sight of each student

The rapid pace of activities leaves some students behind

Figure 5.3 Equitable Access Target 2, addressing student needs.

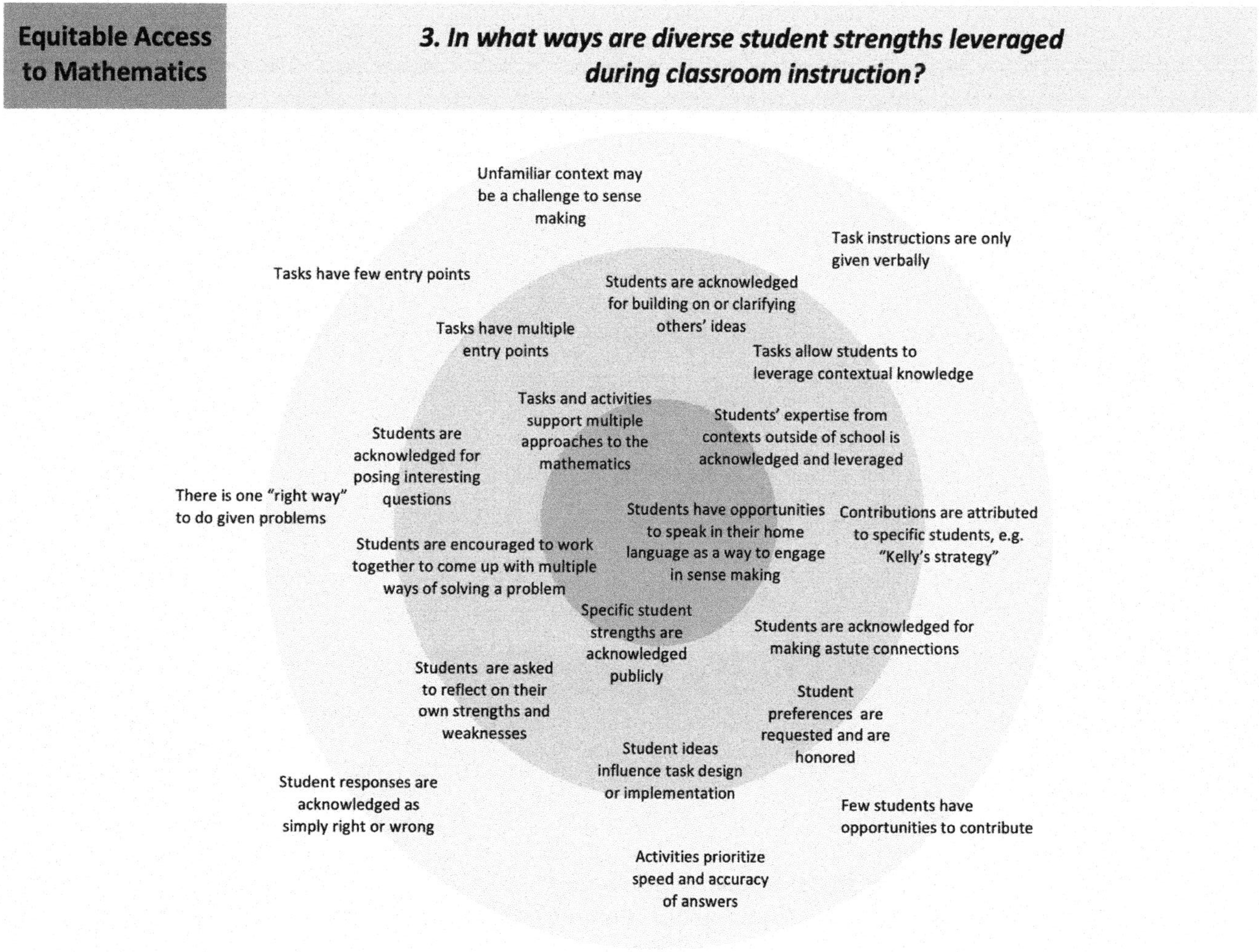

Equitable Access to Mathematics

3. In what ways are diverse student strengths leveraged during classroom instruction?

Task instructions are only given verbally

Students are acknowledged for building on or clarifying others' ideas

Tasks allow students to leverage contextual knowledge

Students' expertise from contexts outside of school is acknowledged and leveraged

Contributions are attributed to specific students, e.g. "Kelly's strategy"

Students have opportunities to speak in their home language as a way to engage in sense making

Students are acknowledged for making astute connections

Student preferences are requested and are honored

Few students have opportunities to contribute

Unfamiliar context may be a challenge to sense making

Tasks have multiple entry points

Tasks and activities support multiple approaches to the mathematics

Specific student strengths are acknowledged publicly

Student ideas influence task design or implementation

Activities prioritize speed and accuracy of answers

Tasks have few entry points

Students are acknowledged for posing interesting questions

Students are encouraged to work together to come up with multiple ways of solving a problem

Students are asked to reflect on their own strengths and weaknesses

There is one "right way" to do given problems

Student responses are acknowledged as simply right or wrong

Figure 5.4 Equitable Access Target 3, leveraging student strengths.

Equitable Access to Mathematics

To what extent are all students provided opportunities to engage with the core content and practices of the lesson?

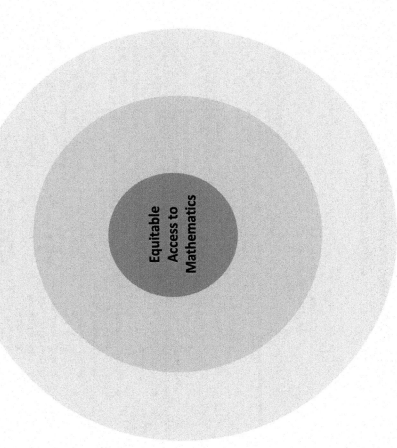

Build your own!

Equitable Access to Mathematics

Figure 5.5 Make your own equitable access target!

Using the equitable access targets – three examples

Here are our equitable access examples, in ascending grade order. Each example features one strategy.

An elementary grades example – multiplication and division word problems

Task enhancement strategy: three reads and modification of task

Students are introduced to multiplication and division in various ways across the elementary grades. By fourth grade, they encounter contextual situations in which modeling multiplication and division is central. Figure 5.6 shows the text of a grade 4 worksheet taken from the web. The worksheet was designed for individual practice or for homework.

Broadly speaking, this task is about modeling and the different kinds of situations in which multiplication, division, proportional relationships, and remainders are all relevant and meaningful in such situations. Let's look at the targets for Dimension 3 (Equitable Access) as we think about using this task in the classroom.

Identifying the challenges

On Target 1:

– Aspects of the context may be unfamiliar.
– The task is linguistically complex, without sufficient supports for students to make sense of this complexity.

Multiplication and division word problems
Grade 4 Word problems worksheet

A hotel has 7 floors. The lobby, restaurant, and gym are located on the ground floor. The guest rooms are on the 1st to 6th floors.

1. If there are 35 standard rooms on each floor, how many standard rooms are there?
2. There are 4 housekeepers working on each floor. One room only requires one housekeeper. If the housekeepers try distributing the work equally, how many housekeepers need to clean more rooms than the others?
3. If each standard room can fit 2 guests, what is the maximum number of guests that all the standard rooms can accommodate?

Figure 5.6 A word problem in a fourth grade worksheet from K5learning.com. Reproduced with permission.

On Target 2:

– High linguistic threshold prevents some students from engaging.
– Students are provided little support for significant challenges.

On Target 3:

– Unfamiliar context may be a challenge to sense-making.

Identifying the opportunities

The inner rings of the three Equitable Access targets suggest various opportunities to open the task up to more students.

On Target 1:

– Pre-reads and other strategies make sure all students understand what problems ask of them.
– Tasks can be approached in multiple ways.

On Target 2:

– Activities provide support for emerging bilingual students' learning needs.

On Target 3:

– Students' expertise from contexts outside of school is acknowledged and leveraged.
– Students have opportunities to speak in their home language as a way to engage in sense-making.

Generating possibilities

The equitable access dimension of TRU centers the question: *To what extent are all students provided opportunities to engage with the core content and practices of the lesson?* This speaks not only to the way the task is framed but also to task-related interactions that occur during the lesson.

Students may know what a hotel is and what it looks like. However, the language in the task prompt is complex, and some students may not be familiar enough with the context to understand what the task calls for. For example, problem 2 is a "remainder" problem, but it takes work to see it that way. One move that the teacher can make is to provide students with pre-reading strategies that support all students, especially emergent bilinguals.

The three reads protocol (a common literacy strategy; see Zwiers et al., 2017) involves reading a math scenario three times with a different goal each time. The first read is aimed at understanding the context. The second read is aimed at understanding the mathematics involved. The third read is aimed at generating a plan for solving the problem. This protocol can be used individually, with small groups or the whole class. It provides opportunities for students to build on each other's understandings. Using the three reads strategy can level the playing field, supporting all students' access to the main ideas in the task.

We offer two other possible task modifications. Here's a modification of the first question: *If there are 35 standard rooms on each floor, write a number sentence that describes the total number of rooms in the hotel. Show two different models that explain how to calculate the product.* Students who are familiar with different ways to model multiplication may use tape diagrams, area models, arrays, and number lines. Posing the question this way can open opportunities for students to see the benefits of using various models for a given scenario and can create opportunities for building connections among various models. The more detailed representations may also help clarify why question 2 is a "remainder" problem.

Second, we might change question 3 to use children's knowledge from outside the classroom. Consider the following: *You can design a floor in the hotel with different sized rooms for your family and friends, with one restriction – you can have at most ten rooms. If all the rooms are full, how many occupants are in your floor? Can you explain how you came up with that number?* This problem variant can still lead to complex mathematical situations that need to be worked through – but this time the children are in charge of them. They may decide the hotel will have large rooms and small rooms, based on their family size. This kind of problem not only leverages students' everyday knowledge but also creates opportunities for students to take ownership of their design and to show multiple ways of modeling multiplication problems. In addition, discussing how many people will fit into (say) 3 rooms that hold 1 person, 3 rooms that hold 2 people, and 5 rooms that hold 3 people raises interesting issues about subtotals, strategies for getting the answer, and more.

A middle grades example – the tile pattern challenge

Task enhancement strategy: highlight connections between multiple representations

In middle school students are introduced to various representations of linear functions, including equations, graphs, and tables. Students are taught how to substitute various values of x into an equation to produce a table of values, which can then be used to plot points on a line. Students practice these skills by doing collections of exercises like the ones shown in Figure 5.7.

Complete the function table. Plot the points and graph the line.

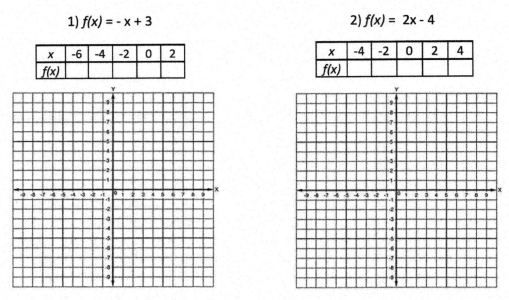

Figure 5.7 The first two (of many!) practice problems on calculating functions and plotting graphs.

Lessons on graphing linear functions eventually lead to lessons on slope. Students learn that slope equals rise over run or change in y over change in x.

Identifying the challenges

On Target 1:

- Activities are limited to only one "right way" to do the given task.

On Target 2:

- Mistakes are corrected at the procedural level, not addressing underlying issues.

On Target 3:

- Activities prioritize speed and accuracy of answers.

Identifying the opportunities

On Target 1:

- Activities supported forms of participation (speaking, writing, listening, demonstrating, etc.).

On Target 2:

– Scaffolding allows each student to engage in rich mathematics.

On Target 3:

– Tasks and activities support multiple approaches to the mathematics.

Generating possibilities

The tasks presented in Figure 5.7 are intended to help students move from one representation of a linear function to another. However, the repetitive nature of such problems, the presumed independence of task completion, and the relatively narrow focus on skills limits the access students have to rich mathematical engagement. Some students might breeze through the procedures quickly, filling in the tables and plotting points without much thought. Other students who had not yet mastered these procedural skills could get stuck, not knowing how to proceed toward the correct answers without guidance from their teacher.

Tasks centered on multiple representations of functions offer a range of opportunities to expand equitable access to important mathematics for students. A classic CPM example is presented in Figure 5.8.[2]

This task begins with a concrete representation of the tile pattern. It then guides students toward abstraction (writing a rule). The focus on connections between representations pushes students past procedural skills into more conceptual understandings of growth patterns. Implementing the task as a group challenge opens up opportunities for more students to share their thinking by explaining patterns they see in the growth of the figures and by

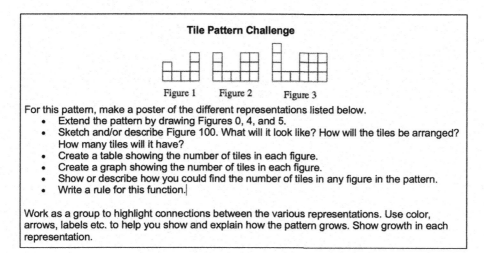

Figure 5.8 The tile pattern challenge. Reproduced with permission from CPM.

demonstrating their understandings in multiple ways. The various representations serve as scaffolding for students as they take steps toward more formalized and more generalized rules. The richness here lies in making sense of the mathematics, specifically how the representations are connected to one another, not simply in carrying out procedures.

A secondary grades example – solving linear equations

Task enhancement strategy: use discourse routines, specifically "Sage and Scribe"

Solving linear equations is an important topic that often warrants practice. A typical worksheet is shown in Figure 5.9.

Identifying the challenges

On Target 1:

−   Student work is mostly individual.

On Target 2:

−   Mistakes are corrected at the procedural level, not addressing underlying issues.

On Target 3:

−   Tasks have few entry points.
−   Student responses are acknowledged as simply right or wrong.

Multi-Step Equations
Solve each equation.

1) $-15 = -2a + 5a$ 2) $5 = 2 - 3b + 6$

3) $9c - 4 = 2 - 2c$ 4) $3d + 6 = 5d - 12$

5) $4e - 2e + 1 = 3 - 4e$ 6) $1 - f = 3f + 17$

7) $7g - g + 3 = 3g$ 8) $5h + 2h - 4 = 6 - 7h$

Figure 5.9 A worksheet for practicing linear equation solving.

Identifying the opportunities

On Target 1:

–   Every student has opportunities to speak and to listen.

On Target 2:

–   Techniques like "think/pair/share" provide each student time and space for meaningful engagement.
–   There is time for students to work through ideas.
–   The environment provides resources to help students when they get stuck.

On Target 3:

–   Activities provide support for emerging bilingual students' learning needs.

Generating possibilities

The targets highlight several issues related to pacing and student support. Most of the curricula we know, even those that are problem-based, include a good number of practice problems solving linear equations. That's fine for skill development. Yet, there are real concerns that students will either speed through the worksheet (if they are already confident with these problems) or give up quickly (if they aren't) and won't discuss their reasoning enough to advance their thinking. Access and mathematical richness are likely to improve if students have more structured opportunities to talk about their strategies, explain their thinking clearly enough to support learning, and persevere because they are able to identify what they do and don't know and get help.

One strategy is to use the same worksheet along with a discussion routine. In the "Sage and Scribe" routine, two students complete one worksheet together, taking turns in different roles as they solve the problems jointly. One student, the "sage," explains verbally how to solve the first problem while the other student, the "scribe," writes down the work and asks questions as needed. The students then switch roles for the next problem. This routine forces the "sage" to explain their process clearly enough for the "scribe" to understand while offering both students many opportunities to help each other and ask questions. To support emerging bilinguals (and all students) in the mathematical language of the activity, we can create sentence starters and have one or two pairs of students model the sage and scribe routine at the front of the class during the launch.

Notes

1   This example comes from the Mathematics Assessment Project's formative Assessment Lesson "Evaluating statements about length and area." The lesson and 99 others can be downloaded at no cost from http://map. mathshell.org/lessons.php.
2   This task was adapted from CPM curriculum (cpm.org), at https:// slideplayer.com/slide/8978926/.

6

Dimension 4
Agency, Ownership, and Identity

What the agency, ownership, and identity dimension involves

Are you a reader? A sports fan? A history buff? Or…?

Whatever our preferences and passions, we talk about what we like. We seek out opportunities to engage with the things we enjoy, and we dig more deeply into them. They're part of our identity. Your mathematical identity includes how you see yourself in relation to mathematics. Do you think you're good at math? Do you like it? Are you willing to jump in and try things? Or does math make you uncomfortable and do you shy away from it when you can?

If you see yourself as a mathematical sense-maker (a.k.a. a "math person"), then you're willing to dig into math problems, even when they're challenging. You enjoy mathematical conversations. You feel a sense of accomplishment when you make progress, by yourself and with others; you "own" the results because you've figured things out in ways that make sense to you. Being someone who enjoys mathematics is part of your identity.

One goal of mathematics instruction is to support every student to see themself as a mathematical sense-maker. These three aspects of being a math person – mathematical agency, ownership, and identity – are critically important.

A student's mathematical agency involves actively participating in mathematical conversations, taking intellectual risks and sharing ideas. It's a willingness to take initiative in mathematics.

DOI: 10.4324/9781003376903-6

> *A student's ownership of mathematics comes from that student's working through ideas, individually or collectively, and making those ideas their own – the sense that "I (or we) figured this out; it makes sense to me; it's not simply what 'they' told me is true."*
>
> *A student's mathematical identity refers to who they see themself to be in relation to mathematics. A student with a positive mathematics identity is likely to think, "I like math; it makes sense; and I can figure things out." Math is part of themself, in and out of school.*

Students develop these attributes from their experiences participating meaningfully in mathematical activities. A sense of initiative comes from having opportunities to think things through, volunteer ideas, work with others' ideas, and have one's own ideas taken up and worked with. A sense of authority comes from having "authored" ideas. A sense of ownership over ideas comes from having done some real sense-making.

It's essential to note that all of these opportunities are *social* opportunities – identities are socially constructed. The way we see ourselves is fundamentally shaped, first, by the ways that we see ourselves fitting into various environments (do we feel welcomed?) and second, by the ways that others interact with us (in academic language, how others position us). If we feel at home and our ideas are treated with respect, that shapes our identities in positive ways. If we don't see ourselves reflected in the environment or we are positioned as not having much to offer it, our identities are shaped in negative ways.

People's disciplinary identities are shaped by their experiences both outside and inside school. In math, stereotypes about gender, race, and ethnicity play a significant role. Common narratives such as "girls are bad at math," "Asians are good at math," and "Black and Latinx students are bad at math" shape both how people see themselves and not only how others see them, but how they act toward them (Martin, 2009; Nasir & Shah, 2011). These biases shape people's mathematical senses of self and their actions, even before – and as – they enter the classroom. For example, Shah (2017) relates the story of an African American student who was asked "what are you doing here?" as he entered a high school calculus class that was otherwise populated by white and Asian American students. No matter the demographic distribution of students in a school, societal myths about different groups' mathematical abilities permeate school walls and need to be addressed in classrooms.

It follows that our classrooms need to be welcoming to students with a wide variety of mathematical and cultural backgrounds, formal and informal knowledge, strengths and challenges, language repertoires and fluency, etc. Our classrooms need to provide every student with opportunities to participate meaningfully and productively in the mathematical work of the class. Dimension 4 is about the nature of the classroom as a welcoming mathematical community.

This dimension draws heavily on the wealth of literature about creating inclusive learning spaces, with explicit attention to issues of race, culture, and power. Culturally relevant pedagogy (CRP), pioneered by Gloria Ladson-Billings (1994, 1995), includes attention to (a) academic success and high expectations, (b) cultural affirmation and the development of cross-cultural competence, and (c) sociopolitical critique. CRP supports students' cultural and disciplinary identity development through attention to issues of curriculum, cultural repertoires of practice for participation (Gutierrez & Rogoff, 2003), and issues of inclusion and exclusion in classroom social spaces.

A bit more background

As emphasized above, the development of identity is a fundamentally social process. Developing and refining your understandings, picking up and refining practices, and being introduced to alternative perspectives or nuances that might have skipped your notice will all come from engaged interactions with others – if you feel welcomed into the relevant community.

Identity involves a sense of engagement with whatever it is that matters to you and a sense of community. It doesn't matter which part of your identity you focus on, whether it's being a family member, literature fan, movie buff, or some aspect of your professional work – it's in contexts where you engage deeply that you develop those aspects of your identity.

Your engagement involves the *practices* associated with the relevant domain and the active use of knowledge. Yes, sports fans may know batting averages, free throw or passing percentages, point spreads and playoff records; but they also know sport strategy and are happy to debate the merits of a decision made by a coach or manager. Foodies do the same. There are writing groups, in which members exchange drafts and ideas as they refine their craft. Math class should be no different (see Gutierrez & Rogoff, 2003).

The ongoing development of identity is a matter of *personal history*. How you see yourself in any endeavor depends on not simply past successes and failures but on how you've been welcomed into the community, what kinds of feedback and encouragement you've received, and so on. We're focused on what happens inside classrooms, but we also have to recognize that societal biases about who's good or bad at math permeate classroom doors, that patterns of tracking at the school or district level often reinforce those biases, and that classroom interactions can do so as well.

What happens in classrooms is critical for identity development, because identity is dynamic. Positive experiences and inclusive communities can kindle interests and engagement that were previously latent or even discouraged. Here is one example.

What would you do as a teacher if a young Black woman with an IEP walks into your classroom in her first week of ninth grade and says "this class makes me feel dumb"? This student has barely experienced your class. It's likely previous negative experiences have shaped how she feels walking into your class. One of the authors experienced this situation and responded by reaching out to the student through one-on-one check-ins, positive reinforcement, consistent communication and collaboration with her mom, and supporting her during class – all while consistently treating the student as a knowledgeable and capable learner whose ideas and questions made sense. By the end of the first month of school, this student explained her mathematical thinking at the front of the class. The young woman learned important mathematics in this class. At least as important, she learned that she was capable of doing challenging mathematics.

Three related ideas are useful to consider.

The first concerns the relationship between mathematics and people's lived experiences. Most people experience mathematics as an abstract discipline that has little to do with either the "real world" or their personal lives; when they enter math class they enter a separate universe in which much of their knowledge is of little if any value. The unfortunate consequence of this kind of experience is that it's hard for many people to connect to mathematics as something personally meaningful.

Things don't have to be this way. Consider curricular materials, for example. There are many possibilities for classroom activities that build on funds of knowledge in nondominant communities (Moll, Amanti, Neff, & Gonzalez, 1992). There is a closely related literature on mathematics for social justice, where issues related to students' personal and cultural experiences often provide the contexts for mathematical explorations (see, e.g., Bartell, Berry, Felton-Koestler & Yeh, 2020; Berry, Conway, Lawler & Staley, 2020; Conway et al., in press; Gutstein & Peterson, 2013). While our emphasis here is on classroom practices, these issues open the door to much larger issues of social justice. See, for example, the catalog at https://rethinkingschools.org/. See also Schoenfeld and Brown (in press). A broad range of topic areas is ripe for exploration, e.g., issues related to Covid-19 (vaccination, masking, and social distancing), modeling phenomena such as the census and redistricting, and "mathematical literacy" more generally (see, e.g., Burkhardt & Schoenfeld, 2022). To the degree that we can incorporate topics into the curriculum that students can see as relevant and meaningful, we open the door to personal connections, agency, and identity.

A second related idea is the concept of *authority*, the idea that active participants in mathematical communities author mathematics and make it their own. In a paper that identifies many useful resources, Engle (2011) argues that four kinds of intellectual authority are important and synergistic: intellectual agency, authorship, contributorship, and being positioned as a local authority. We note that the more that students can draw upon the wide range of resources potentially available to them (including cultural funds of knowledge and language repertoires), the more ideas they can bring to mathematical conversations, and the more ownership they will have over those ideas, and the more of a sense of mathematical authority they will develop.

A third related idea is the notion of growth mindset (Dweck, 2007, 2012). Many people in the US tend to believe that mathematical ability is innate and fixed, that you're either born with mathematical aptitude or you're not, and that nothing can change that state of affairs. An unfortunate consequence of this fixed mindset is that people who believe that they don't have mathematical talent are likely to give up on mathematics. In contrast, having a growth mindset involves the (empirically valid) assumption that people's mathematical performance gets better if they work at it. Whether or not one's mathematics teaching attends formally to issues of mindset, it's good to remember that an environment that supports engagement and builds on partial success will result, over time, in people coming to see themselves increasingly as mathematical sense-makers (see, e.g., Boaler & Staples, 2008; Nasir, Cabana, Shreve, Woodbury & Louie, 2014).

Addressing such issues will enrich mathematics curriculum and instruction.

The overarching question for Dimension 4 (Agency, Ownership, and Identity, or AOI) is: *How do classroom activities invite students to connect to mathematics – to explore, to conjecture, to reason, to explain, and to build on emerging ideas, helping them develop agency, personal ownership of the content, and positive disciplinary identities?*

Three key questions, elaborated in the targets that conclude this dimension, are as follows:

- How do classroom activities help students connect their personal and cultural identities with their mathematical experiences, so they can see themselves as mathematical sense-makers?
- In what ways do classroom activities support students in building on the resources they bring to the classroom, taking intellectual risks, and sharing their developing thinking?
- In what ways do classroom activities provide opportunities for students to make the mathematics their own?

Some ideas that may help

In general, the ways that we open up mathematical activities, such as supporting multiple pathways into mathematical problems and providing opportunities for sense-making, including drawing on students' personal and cultural knowledge, help to provide more fertile ground for AOI.

Equitable Access to Mathematics and AOI go hand in hand. Every student should have meaningful opportunities to develop a positive mathematical identity by engaging in meaningful ways with core mathematical content and practices, and that means they need access to full participation in core mathematical activities. Specifically, some of the following should be part of our AOI toolkit:

Establishing and maintaining classroom norms that create safe spaces for academic risk-taking.

When students feel free and safe to contribute, they develop a sense of agency and the class profits from their ideas. When they don't the students lose opportunity for positive mathematics identity development and the class is deprived of their ideas.

Encouraging and supporting mathematical contributions by each and every student.

The discussion of equitable access highlighted participation tools and structures such as *think/pair/share, whip-arounds, sentence starters/frames, participation chips,* and *gallery walks.* All of them can be used in ways that open AOI. Here is a story from a kindergarten classroom.

The teacher had posed a challenging problem, and students were working on it at their tables in groups of two or three. The teacher circulated through the classroom, asking questions and providing clarifications – not answers or directions that took away the challenge – when students were having difficulties. Graciela (a pseudonym) was having a hard time. She had misinterpreted the task and it had taken her a while to understand the it. But then she worked slowly and carefully to figure things out.

When the teacher brought the students together to discuss the problem, he asked Graciela to come to the front of the room to discuss what she had done. She explained what she had done and then, with some scaffolding from the teacher, fielded questions about how she had worked through the problem. The teacher asked who understood what she had done, and most but not all hands went up. The teacher scaffolded conversations between Graciela and those who hadn't followed, until all the students said they understood her solution.

The teacher then asked Graciela, "Was it easy for you to solve this problem?" She answered "No, when I started I didn't understand it. Then I made some mistakes and it took me a long time." The teacher asked, "Are you sure you got it right?" Graciela said yes.

The teacher asked, "Does everybody understand what Graciela did?" There was a chorus of yeses. He continued, "You know, mathematicians, Graciela showed us what it's like to be a good mathematician. Sometimes you don't understand a problem right away. Sometimes you don't know what to do to get started, and you have to work really hard. Math can be hard. But if you work at it, then you'll get it. That's what being a mathematician is. Thank you for showing us, Graciela."

It's important that Graciela was chosen to present her solution. Graciela was positioned, legitimately, as a smart student who had solved a challenging problem. She was also positioned as a model mathematician – as someone who insists on things making sense and who perseveres until she has made sense of things. The discussion contributed to her sense of agency and to the development of her identity as a "math person." Discussions such as this do more than help the individuals who are their focus. They help to build a community of mathematical sense-makers – people who develop a sense of agency, insist that they make sense of the math so that they own it, and have productive mathematical identities.

Supporting broad notions of mathematical competence.

Sense-making, perseverance, and resilience were all seen as aspects of productive mathematical behavior in Graciela's classroom. There are many more. Some students are good at asking questions, prompting everyone to think more deeply. Some are good at reflecting on what the group has done and making sure that it fits together. Some are good at being organized. Some are good at cleaning up mathematical arguments or seeing connections. All of these ways of thinking mathematically contribute to individual and collective understanding. When students come to see them as such, they have more ways to participate in the class's mathematical activity. This means more opportunities for AOI. (Recall Complex instruction, e.g., Horn 2007, Cohen & Lotan, 1997.)

Opening up room for students to feel welcomed by mathematics and for them to bring their broad range of informal, cultural, and linguistic resources into mathematical conversations.

The more that math relates in meaningful ways to issues in students' lives, the more they are likely to see themselves in mathematics – and to experience and appreciate the power of mathematical thinking. Two relevant resources are *Rethinking Mathematics* (Gutstein & Peterson, 2013) and the Corwin series on mathematics and social injustice (e.g., Bartell et al., 2020; Berry et. al., 2020).

Providing support as needed, to remove barriers to participation.

All too often, "scaffolding" means "making things easy," with the result being that some of the challenges necessary for mathematical growth are removed. The scaffolding for Graciela ended when she was in a position to grapple fully with the problem – when she was in a position to do mathematical sense-making. Many of the techniques discussed in Dimension 3, e.g., "three reads," "collect and display," "critique, correct, and clarify," and "stronger and clearer each time," help position students so that they can engage in mathematical sense-making.

Building classroom norms that focus on the practices of doing mathematics rather than "answer-getting."

In many classes, work on a task is considered done when someone shows "the answer." Questions such as "how do we know this is true?" and "did anyone do this a different way?" can keep the focus on mathematical thinking.

Understanding that errors are a natural part of sense-making.

Student suggestions, even when incorrect, often contain the germs of useful ideas. Creating a climate in which it is safe to put forth partially formed ideas and work through them creates a safe climate – and one that can be mathematically richer, as ideas are explored.

Having students present their thinking and revoicing student contributions in ways that leave ownership with the students.

In the example above, Graciela had center stage and the teacher a supporting role.

Attributing ownership of the ideas to the students who created them.

Referring to the work students author by name, e.g. "Graciela's method," helps strengthen students' mathematical ownership and identity. Note that we need to be careful to equitably distribute opportunities for students to present and/or to have their ideas highlighted!

Maintaining a balance of challenge and support is essential. A real sense of success comes from having faced a challenge and overcome it. This means that students must be given room to struggle (Dimension 2: Cognitive Demand) and that the degree of struggle and frustration must be monitored so that the possibility of meaningful success is within reach (Dimension 5: Formative Assessment).

A second classroom episode points to the importance of small group interactions in shaping AOI and the interactions of AOI with cognitive demand. A full analysis of the episode, "Where is the ten," can be found in Schoenfeld et al. (2023). In the classroom, which employs Complex Instruction, one student in a small group has been designated as the group's "explainer," which means that she is responsible for explaining the group's solution to a problem to the teacher. Her groupmates are responsible for supporting her understanding. The problem is challenging, and the student's frustration is evident at times – but she and her groupmates work through the problem and finally meet the teacher's very high standards for explanation. The result is far more than relief for the focal student – it's empowerment. Having faced a meaningful challenge and overcome it, she is more confident and begins the next task with visibly more initiative and agency.

The moves highlighted above reward student agency, promote ownership, and help to build positive mathematical identities. Generally speaking, research suggests that effective teachers recognize and capitalize on the strengths of individual students, using varied approaches to help students establish footholds in the learning community (Boaler, 2008; Horn, 2007, 2012; Cohen & Lotan, 1997). This can be done by publicly recognizing and reinforcing the strengths and abilities of individual students – by, for example, attributing competence to students in class discussion and in group work, as in the vignettes above, re-engaging students who are struggling, and challenging individual students to elaborate on their own ideas and the ideas of their classmates. See Michaels, O'Connor, and Resnick (2010) for discussions of these and other discourse moves in the service of "accountable talk," in which students take increasing responsibility (and credit!) for working through mathematical ideas.

More specific questions for planning and review

Here are some specific questions you might consider as you work to open up opportunities for AOI. The core questions are expansions of those from the *TRU Math Conversation Guide* (Baldinger, Louie, and the Algebra Teaching Study and Mathematics Assessment Project, 2018).

- Who generates the ideas that get discussed?
    - Are there ways to open things up – perhaps, again, by using participation tools and structures such as think/pair/share, whip-arounds, sentence starters, participation chips, and gallery walks?
- In what ways do curricular or classroom activities invite students into mathematics, providing opportunities for students to see mathematics as something meaningful in their own lives?
- What kinds of ideas do students have opportunities to generate and share (strategies, connections, conjectures, partial understandings, prior knowledge, and representations)?
    - Are there ways to validate the various ways of "thinking mathematically," so that the range of ways students can contribute is expanded?
- How might we create more opportunities for more students to see themselves and each other as powerful mathematical thinkers?
- What gets validated as a good contribution?
- Which students get to explain their own ideas? To respond to others' ideas in meaningful ways?
- Who evaluates and/or responds to others' ideas?
    - Are there ways to "deflect authority" from the teacher and to "spread the wealth" equitably among students?
- How deeply do students get to explain their ideas?
    - This, of course, is related to Dimension 1: How can we make sure there are opportunities for connections and explanations, and that they are seen as part of what it means to engage in mathematics?
- How does (or how could) the teacher respond to student ideas (evaluating, questioning, probing, soliciting responses from other students, etc.)?
    - What ways are there for shaping the classroom community as a collective, so that students see themselves, individually and collectively, as sense-makers? Taking stock:
- How are norms about students' and teachers' roles in generating ideas developing?
- How are norms about what counts as mathematical activity (justifying, experimenting, connecting, practicing, memorizing, etc.) developing?

– Which students seem to see themselves as powerful mathematical thinkers right now?

Some resources that might be useful

Appendix A contains a collection of classroom strategies that serve multiple purposes. Specific strategies that are often useful for opening up mathematical opportunities for agency, ownership, and identity include *What's the big picture, Share what your partner said, Name it, Confidence thermometers, Time alone, Student-led questioning routines, Choose your color, Start with the end, and Turn it back to students*. Many of the texts referenced above, e.g., Cohen & Lotan (1997, 2014), Gutstein and Peterson (2013), Horn (2012), Mason, Burton and Stacey (2010), Nasir et al. (2014) and Zwiers et al. (2017) contain practical suggestions, as do the TRU Math *Conversation Guide* (Baldinger et al., 2018) and the TRU Math *Observation Guide* (Schoenfeld et al., 2018). The remaining references are well worth reading as background.

Further resources

The references in this chapter just touch the tip of the iceberg. Other useful resources can be found in Banks & Banks (2004), Bartell et al. (2020), Berry et al. (2020), Boaler & Staples (2008), Brown et al. (1993), Darling Hammond (2010), Esmonde (2009), Gresalfi, Martin, Hand & Greeno (2009), Martin (2009), Moll et al. (1992), Moschkovich (2012, 2013), Nasir et al. (2014), Nasir & Cobb (2007), and Rosebery, Warren & Conant (1992).

The agency, ownership, and identity targets

Figure 6.1 lists the core questions for the AOI targets; Figures 6.2–6.4 elaborate on those core questions. You may want to articulate your own goals or ideas on the target in Figure 6.5. Take a look at the tasks you're planning to (or did) use. See if you can place them on the targets and if you can modify them so that they move more toward the center.

Again, if you can do this with a colleague – even better, if the colleague can visit your class and take notes – then you can profit from having a spare set of eyes and comparing notes.

Agency, Ownership and Identity

How do classroom activities invite students to connect to mathematics — to explore, to conjecture, to reason, to explain and to build on emerging ideas, helping them develop agency, personal ownership of the content, and positive disciplinary identities?

Core Questions:

1. How do classroom activities help students connect their personal and cultural identities with their mathematical experiences, so they can see themselves as mathematical sense makers?

2. In what ways do classroom activities support students in building on the resources they bring to the classroom, taking intellectual risks, and sharing their developing thinking?

3. In what ways do classroom activities provide opportunities for students to make the mathematics their own?

Figure 6.1 Core questions about Dimension 4, Agency, Ownership, and Identity.

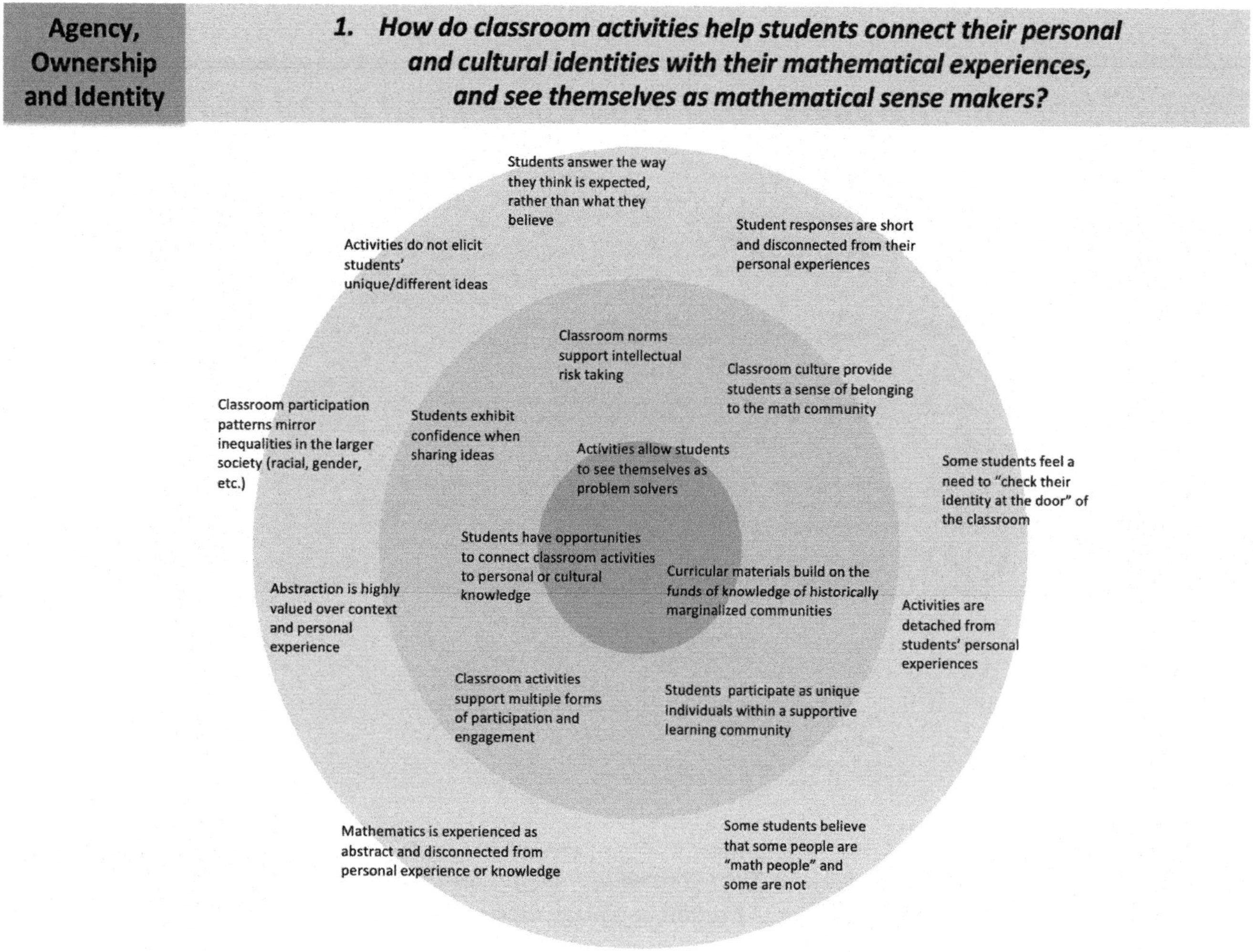

Agency,
Ownership
and Identity

1. How do classroom activities help students connect their personal and cultural identities with their mathematical experiences, and see themselves as mathematical sense makers?

Some students feel a need to "check their identity at the door" of the classroom

Student responses are short and disconnected from their personal experiences

Activities are detached from students' personal experiences

Classroom culture provide students a sense of belonging to the math community

Students answer the way they think is expected, rather than what they believe

Classroom norms support intellectual risk taking

Activities allow students to see themselves as problem solvers

Curricular materials build on the funds of knowledge of historically marginalized communities

Students participate as unique individuals within a supportive learning community

Activities do not elicit students' unique/different ideas

Students exhibit confidence when sharing ideas

Students have opportunities to connect classroom activities to personal or cultural knowledge

Classroom activities support multiple forms of participation and engagement

Some students believe that some people are "math people" and some are not

Classroom participation patterns mirror inequalities in the larger society (racial, gender, etc.)

Abstraction is highly valued over context and personal experience

Mathematics is experienced as abstract and disconnected from personal experience or knowledge

Figure 6.2 AOI Target 1, Connecting to students' personal and cultural identities.

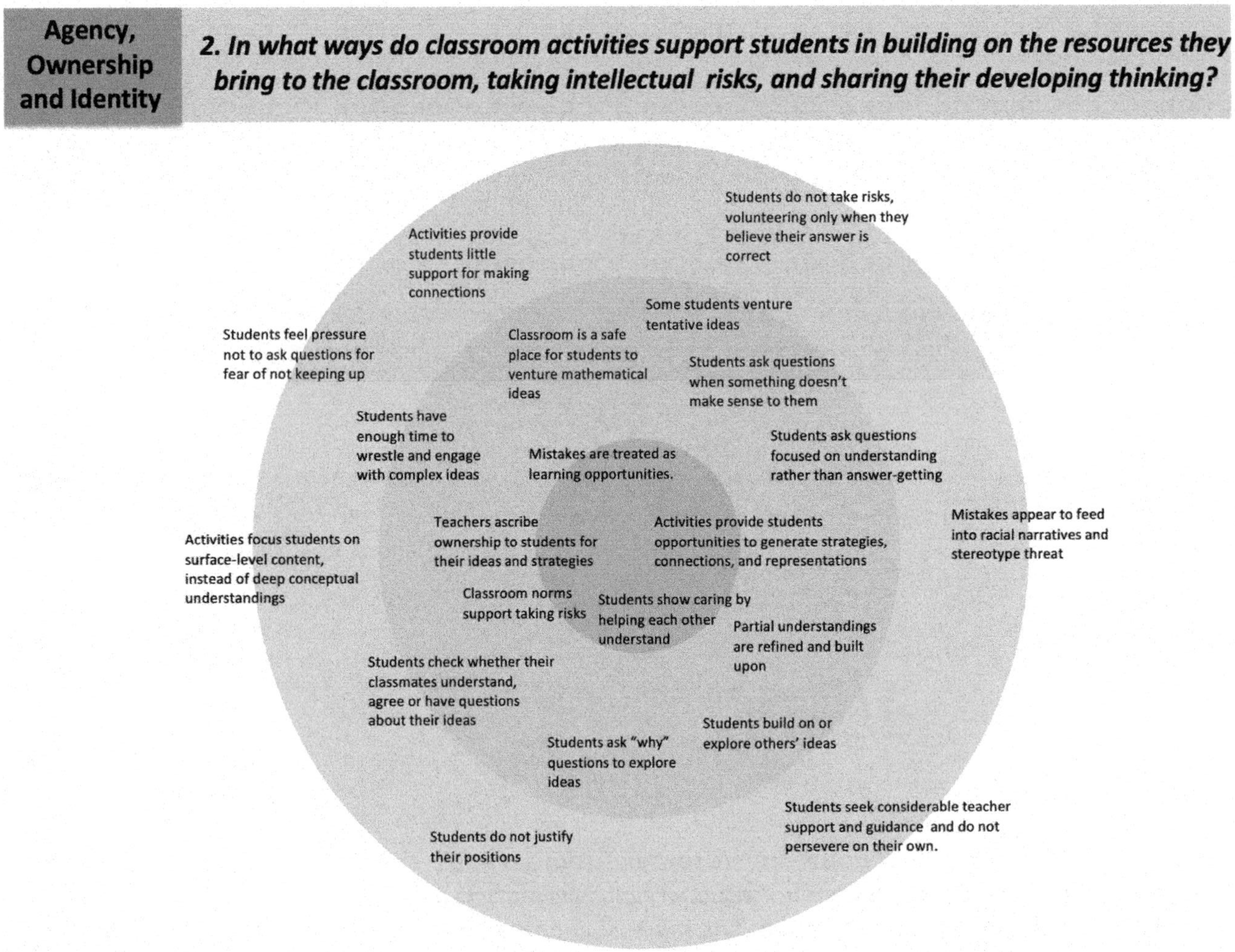

Agency, Ownership and Identity

2. In what ways do classroom activities support students in building on the resources they bring to the classroom, taking intellectual risks, and sharing their developing thinking?

Students do not take risks, volunteering only when they believe their answer is correct

Some students venture tentative ideas

Students ask questions when something doesn't make sense to them

Students ask questions focused on understanding rather than answer-getting

Mistakes appear to feed into racial narratives and stereotype threat

Activities provide students little support for making connections

Classroom is a safe place for students to venture mathematical ideas

Mistakes are treated as learning opportunities.

Activities provide students opportunities to generate strategies, connections, and representations

Partial understandings are refined and built upon

Students build on or explore others' ideas

Students seek considerable teacher support and guidance and do not persevere on their own.

Students feel pressure not to ask questions for fear of not keeping up

Students have enough time to wrestle and engage with complex ideas

Teachers ascribe ownership to students for their ideas and strategies

Students show caring by helping each other understand

Classroom norms support taking risks

Students ask "why" questions to explore ideas

Activities focus students on surface-level content, instead of deep conceptual understandings

Students check whether their classmates understand, agree or have questions about their ideas

Students do not justify their positions

Figure 6.3 AOI Target 2, Building on student resources.

Agency,
Ownership
and Identity

*3. In what ways do classroom activities provide opportunities for students
to make the mathematics their own?*

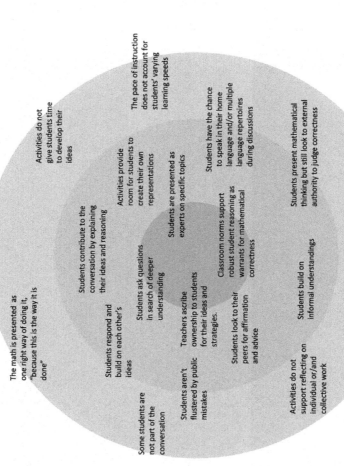

The pace of instruction
does not account for
students' varying
learning speeds

Activities do not
give students time
to develop their
ideas

Students have the chance
to speak in their home
language and/or multiple
language repertoires
during discussions

Activities provide
room for students to
create their own
representations

Students are presented as
experts on specific topics

Students contribute to the
conversation by explaining
their ideas and reasoning

Students present mathematical
thinking but still look to external
authority to judge correctness

Students ask questions
in search of deeper
understanding

Classroom norms support
robust student reasoning as
warrants for mathematical
correctness

The math is presented as
one right way of doing it,
"because this is the way it is
done"

Students respond and
build on each other's
ideas

Teachers ascribe
ownership to students
for their ideas and
strategies.

Students look to their
peers for affirmation
and advice

Students build on
informal understandings

Students' contributions
are arbitrated by
external authority

Some students are
not part of the
conversation

Students aren't
flustered by public
mistakes

Activities do not
support reflecting on
individual or/and
collective work

Figure 6.4 AOI Target 3, Opportunities for students to make the mathematics their own.

| Agency, Ownership and Identity | *In what ways are classroom activities providing students the opportunity to explore, conjecture, reason, explain and build on emerging ideas, contributing to the development of agency, and ownership over the content, resulting in positive disciplinary identities?* |

Feel free to add your own!

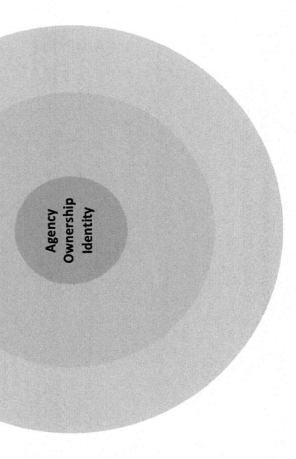

Agency
Ownership
Identity

Figure 6.5 Make your own AOI target!

Using the agency, ownership, and identity targets – three examples

Here are our AOI examples, in ascending grade order.

An elementary grades example – expanding the use of practice problems

Task enhancement strategy: facilitating number talks

Fluency with whole number arithmetic is an important mathematical goal in the elementary grades. Many teachers use worksheets like Figure 6.6 to have students practice two-digit addition. The teacher sets a timer for a few minutes, and students complete as many problems as they can. One unintended consequence of timed work like this is that students are quickly categorized as being "fast" or "slow," with implications for their mathematical identities.

We see two issues with this task. First, the focus on answer-getting masks opportunities for strategic thinking. For example, problems 2 and 6 both have a 9 in the ones digit of the first number. We would hope that at least some students would solve problem 2 by noticing that 19 is one less than 20 and solving the much easier, equivalent problem 20 + 26. A similar strategy would be useful for problem 6, where the student could notice that 99 is one less than 100 and add 100 + 13=113. Without this strategy, problem 6 is quite challenging to solve using a standard algorithm and requires regrouping twice. The activity as presented offers no time for dialogue or reflection about which strategies should be used. In addition to a missed opportunity for rich mathematics, this is a major missed opportunity for students to share *their* strategies and feel a sense of ownership over the mathematics.

Second, asking students to solve many problems under time pressure reinforces the idea that there is one way to solve math problems, whoever is successful at using standard algorithms under time pressure is a "math person," and others are not. This is problematic for students' mathematical identity development.

1)   23
     +16
	2) 19
	     +27
		3) 31
		     +50

4)   38
     +25
	5) 83
	     + 45
		6) 99
		     +14

Figure 6.6 A worksheet for practicing addition.

Identifying the challenges

On Target 1:

– Activities do not elicit students' unique/different ideas.
– Student responses are short and disconnected from their personal experiences.
– Some students believe that some are "math people" and some are not.

On Target 2:

– Activities focus students on surface-level content, instead of deep conceptual understandings.
– Activities provide students little support for making connections.

On Target 3:

– Students' contributions are arbitrated by external authority.

Identifying the opportunities

On Target 1:

– Classroom activities support broad participation.
– Activities allow students to see themselves as problem-solvers.

On Target 2:

– Activities provide students opportunities to generate strategies, connections, and representations.
– Mistakes are treated as learning opportunities.

On Target 3:

– Teachers ascribe ownership to students for their ideas and strategies.
– Classroom norms support robust student reasoning as warrants for mathematical correctness.

Generating possibilities

Number talks (See, e.g., Humphreys & Parker, 2015; Parrish, 2010) are a fluency routine that focuses on strategic thinking, student justification, and building classroom community.

For example, a number talk for problem 2 above could go as follows.

The teacher could begin by writing one of the problems on the board and asking students to think about how to solve it. After giving them time to think the teacher would then ask some students to share their strategies. As they do, the teacher writes each student's name and attributes ownership of the strategy to them. The teacher also often asks questions, uses revoicing, and writes down student thinking to help clarify the student strategy, while always checking with the student whether the teacher's description of *the student's* strategy is accurate. The teacher can also facilitate discussion about which strategy is the easiest or hardest for this problem or might be helpful for other kinds of problems in the future.

Over time, a daily number talk routine gives students substantial fluency practice. Instead of having students use worksheets, practicing a standard algorithm and having their answers arbitrated by an external authority, students justify their ideas, consider others' ideas, make connections, and think strategically about when different procedures are most useful. In addition to enriching the mathematics, number talks substantially strengthen AOI.

A middle grades example – creating problem-solving opportunities

Task enhancement strategy: students create tasks, to fostering ownership and problem-solving

In middle school, students analyze two- and three-dimensional figures using distance, angle, similarity, and congruence. In particular, students work with geometric transformations and their relationship to the concept of congruence. They begin to conceptualize the properties of rotations, reflections, and translations, developing an understanding that these transformations leave the shapes of the initial objects unchanged. Students also come to understand that congruent shapes can be "mapped" one onto the other by using rotations, reflections, or translations.

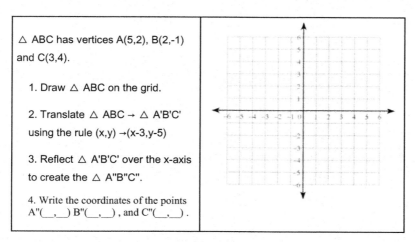

Figure 6.7 A typical geometric transformation task.

A typical example, shown in Figure 6.7, asks students to translate a given triangle and then rotate it over the x-axis.

The next example calls for rotating a triangle 180° around the origin and then reflecting the result over the line $y = x$. Examples continue in this vein.

Once again, the concepts are important and practice is useful. However, the tasks are mechanical in nature and there are no "hooks" to gain the students' interest, much less to foster their ownership of the material.

Here's what we see in the AOI targets.

Identifying the challenges

On Target 1:

–   Activities do not elicit students' unique/different ideas.

On Target 2:

–   Students do not justify their positions.

On Target 3:

–   Activities do not give students time to develop their ideas.

Identifying the opportunities

Here are some possible opportunities.

On Target 1:

–   Activities allow students to see themselves as problem-solvers.

On Target 2:

–   Activities provide students opportunities to generate strategies, connections, and representations.

On Target 3:

–   Students are presented as experts on specific topics.

Generating possibilities

Since there is nothing "special" about the figure to be translated, why not start by asking the students to draw a shape and then choose two transformations that will be applied to that shape?

Their task is to specify each of the transformations using academic language (i.e. "rotate the figure 90 degrees counterclockwise around the origin") and then to perform the transformations. They should identify both the placement of their original shape and their final shape after the transformations.

Once the students have done this, then they pair up to do some problem-solving. Each student gives their partner the coordinates of their original and final shapes. It's the partner's job to find two transformations that will move the initial shape to the final shape.

This is a real problem-solving challenge, in which each student has some expertise and ownership over the figures and transformations they've created and has to solve the problem posed by their partner. Since there can be more than one way to transform one figure to another, there is a lot of room for discovery as well as for working together to troubleshoot attempts that look like they should work but don't. These activities allow students to see themselves as problem-solvers and to develop ownership over the mathematics they have co-created.

A secondary grades example – housing price scenarios

Task enhancement strategy: connect to community issues/concerns

Here is a problem like many found in texts and inline, in which students are asked to create an equation for an exponential growth situation.

> *Carla writes a series of books. She earned $50,000 for the first book, and her cumulative earnings double with each sequel that she writes. Write a function that gives Carla's cumulative earnings, E(n), in dollars, when she has written n sequels.*

This problem can be answered by creating the exponential equation $E(n) = 50,000 \cdot 2^n$, where 50,000 represents the initial earnings and the expression 2^n represents how her cumulative earnings double with each sequel.

Identifying the challenges

This word problem, like many "applications," is artificial, contrived, and unrealistic. We see the following issues.

On Target 1:

- Mathematics is experienced as abstract and disconnected from personal experience or knowledge.

On Target 2:

- Activities focus students on surface-level content, instead of deep conceptual understandings.

On Target 3:

- The math is presented as one right way of doing it, "because this is the way it is done."
- Students' contributions are arbitrated by external authority.

Identifying the opportunities

On Target 1:

- Students have the opportunity to connect classroom activities with personal or cultural knowledge.
- Curricular materials build on the funds of knowledge of historically marginalized communities.

On Target 2:

- Students have enough time to wrestle and engage with complex ideas.

On Target 3:

- Students contribute to the conversation by explaining their ideas and reasoning.

Generating possibilities

There are many ways to open up word problems like this. Among other things, we can ask: How can the problem context be made more authentic and relevant, so that students have the opportunity to connect mathematics to their personal and cultural experience? How can the funds of knowledge of historically marginalized communities be recognized and built on in the task design? Once we identify a culturally relevant problem context, how can it be mathematized in ways that support broad participation in mathematically rich problem-solving?

To address these issues, Alyssa Sayavedra and Warren Currie of the Oakland Unified School District created a unit focusing on an exponential modeling task related to the historical context of redlining in the San Francisco Bay Area[1]. This was of significant potential interest to their students, given the history of housing discrimination against Black families in the Bay Area.

The lesson sequence begins with a data exploration of housing prices in students' own neighborhoods and inequalities in housing prices across the U.S. Students are invited to discuss their experiences of housing inequality and their initial thoughts about the causes. They are introduced to redlining as a historical practice that drives neighborhood segregation and housing inequality, which continue to this day. They complete a brief reading and discussion about "Mr. Barnes," adapted from a true story of a Black father navigating housing discrimination to create financial security for his family.

After the data exploration and reading activities, students then spent about one class period considering the exponential modeling problem given in Figure 6.8.

This problem provides opportunities for students' mathematical sense-making using multiple representations, posters, presentations, discussion, and written reflections. The activity draws students' attention to the mathematical and contextual meaning of the initial value and multiplier in an

---

Housing Price Scenarios

Suppose Mr. Barnes had asked your advice about buying a house in 1955.

Scenario 1: Imagine the house is worth $20,000 but its value increases by 10% each year. Create a table, equation and graph that each include the following information:

      a. What will the house be worth after 1 year?
      b. After 2 years?
      c. After 60 years?

Scenario 2: Imagine the house is worth $100,000 but its value increases by 1% each year. Create a table, equation and graph that each include the following information:

      d. What will the house be worth after 1 year?
      e. After 2 years?
      f. After 60 years?

Now create a poster showing your table, equation, and graph for each scenario. Use color, arrows, labels, etc to show as many connections as possible between the table, equation and graph.

With a partner, write a paragraph summarizing your findings and recommendations to Mr. Barnes.

Figure 6.8 An exponential modeling problem grounded in meaningful real-world data.

exponential equation. These concepts are more meaningful and memorable because of the context in which they are presented.

These activities take extra time – about one to two class days – but they set the tone for the entire unit in ways that are powerful for Agency, Ownership, and Identity. The authors note that a discussion of a current and historical social justice context allows students to share their personal and cultural experiences and to connect their mathematics learning in the unit to an authentic context that can be empowering in the sense of both personal success and social justice. These activities also give students an opportunity to build number sense and vocabulary relevant to the problem in a way that is authentic and uses their prior knowledge. Once this common knowledge base about a social justice context is built in the classroom, it can be referenced in various math activities throughout the unit or year.

Generating such robust examples from scratch is a heavy lift. But resources for similar contextual issues are becoming increasingly available (e.g., Bartell et al., 2020; Berry et al., 2020; Gutstein & Peterson 2013). And, more minor modifications to problems in the curriculum can still render those examples more meaningful and relevant. Collaboration in teacher learning communities is a great way to pool resources for both large and small curricular changes.

Note

1   A version of this task is available online, at no cost, as a resource of the English Learner Success Foundation. Here is a temporary link: https://docs.google.com/document/d/1Y_Ds2hI6JXT1pP0a7VAyPcVooi8E41Y0sUj1KD0jCMY/edit.

7

Dimension 5
Formative Assessment

What the formative assessment dimension involves

Formative assessment involves gathering information about students' understandings before and while teaching to modify lessons in ways that help students engage more effectively with mathematics and each other. The idea is to be responsive to individual and collective student needs. Key questions to keep in mind regarding formative assessment are,

> *To what extent is students' mathematical thinking surfaced?*
> *To what extent does instruction build on student ideas?*

As we teach, we typically encounter challenges in student understanding during the give-and-take of classroom interactions. We may anticipate common errors and be prepared to address them. When students work independently or in small groups, we observe the progress they're making and facilitate learning experiences to help them get unstuck. These are all aspects of good teaching, and they're aspects of formative assessment. And there's more.

The idea is to set up instruction so that it addresses anticipated challenges and builds on students' strengths – with the entire environment (the teacher, the curriculum, and other students) providing support and feedback that helps every student learn. This means making student thinking public and

DOI: 10.4324/9781003376903-7

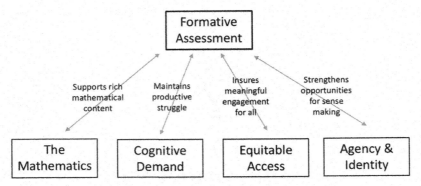

Figure 7.1 The key roles of formative assessment. Reproduced from Burkardt & Schoenfeld (2019, p. 43), with permission.

creating an environment in which students feel safe to voice their ideas, a major feature of AOI. Once student thinking is out in the open, formative assessment can enhance the first four TRU dimensions.

What students reveal about their understanding of mathematics (Dimension 1) allows us to build on what they know and to address gaps or connections we'd like them to make. Having a sense of what students know, in the moment, allows us to observe the level of challenge (Dimension 2). Depending on what we see, we might scaffold, pose an extra challenge, or simply have the students use each other as resources in trying to maintain a reasonable level of cognitive demand. There's also the question of who's engaging, in what ways. Walking through the class, we get a sense of whether there is equitable access to core content and practices – and we try to ensure (Dimension 3) that that's the case. Does every student have the opportunity to get their ideas on the table, to have those ideas discussed, and to do the same for their fellow students? It's from such opportunities for sense-making and the responses to those opportunities that students develop a sense of agency, ownership, and positive mathematical identities (Dimension 4). Formative assessment is the mechanism that enhances this and the other dimensions. See Figure 7.1.

A bit more background

Black and Wiliam (1998, p. 91) originally defined formative assessment as:

> *all those activities undertaken by teachers, and by their students in assessing themselves, which provide information to be used as feedback to modify the teaching and learning activities in which they are engaged. Such assessment becomes "formative assessment" when the evidence is actually used to adapt the teaching work to meet the needs of students.*

This description highlights two essential aspects of formative assessment. First, students play a fundamental role in the process. It's impossible (and not necessarily ideal for student learning) for a teacher to be everywhere at once, addressing all the challenges that emerge. But various activities, including small group discussions in which students are expected to work through their reasoning, explain their ideas carefully, and question the explanations they're given, provide significant opportunities for students to support each other's learning. Such conversations do more than enrich mathematics understanding; they also provide fodder for the development of agency, ownership, and identity. The challenge is to design tasks and activities that bring forth such conversations and to build discourse norms that support them.

Second, the primary purpose of formative assessment is to make productive modifications to instruction. This contrasts strongly with the historical use of the term assessment, a.k.a. "testing." Most testing is conducted with the intention of recording what students have learned. It's *summative*. That's important for grading and bookkeeping purposes, but data indicate that testing after instruction doesn't do much to improve learning. The whole idea is to provide meaningful feedback during the learning process. That's when it makes a difference.

Black and Wiliam (1998) described a series of studies with three conditions regarding the ways homework was returned to students:

1 The papers were returned with comments but no scores.
2 The papers were returned with scores but no comments.
3 The papers were returned with both scores and comments.

Unsurprisingly, returning assignments with comments but without scores (condition 1) significantly improved student learning, while returning assignments with scores but without comments (condition 2) failed to improve learning. Students tend to read and think about comments, but when an assignment containing only scores is returned they glance at their scores and then file the assignment away. And condition 3? When comments *and* scores are returned, that has no more impact on student learning than papers that only have scores. Students look at their grades and, whatever the result, file the papers away.

It's tempting, when you discover a student error, to simply say something like, "That doesn't work, it's better to do it this way." Unfortunately, that approach does not usually help. First, it tends to stifle student agency and to reinforce the message that "things should be done the way the teacher tells you." Second, procedural corrections tend not to work. Incomplete conceptions of mathematics develop over time and are often deeply ingrained in student's understandings. Simply correcting student work leaves very little room for students to develop new understandings in ways that are meaningful and mathematically valid. Thus, it is critical that students be provided opportunities to develop a complete rich network of understandings. (For an early discussion of the resiliency of errors in arithmetic and fractions learning, see Maurer, 1987.)

In general, then, the goal is to structure a sequence of learning activities so that students, individually and with their peers, address learning challenges and are consistently engaged in mathematical sense-making. In discussing design principles for formative assessment, Burkhardt and Schoenfeld (2019) show that formative assessment is facilitated by:

– Using tasks that expose and discuss common partial or incorrect mathematical understandings. (See the lesson "Interpreting Distance-Time Graphs" discussed briefly below for an extended example.)
– Using tasks that provoke cognitive conflicts. There are various ways to do this, e.g.,

  • Asking students to compare their responses with those from other students;
  • Asking students to repeat the task using alternative methods; and
  • Using tasks that contain some form of built-in checks.

  Figure 7.2, from the Formative Assessment Lesson (FAL) "Evaluating statements about length and area" (https://www.map.mathshell.org/lessons.php?unit=9310&collection=8), shows two tasks that stimulate rich conversations. The tasks are challenging, and the requirement that the students carefully justify their answers makes it likely that there will be lively and deeply mathematical conversations.

– Asking students to critique work done by other students. An example from the "length and area" FAL is discussed in Dimension 2.

Some ideas that may help

An illustrative example

To illustrate the suggestions discussed below, we start this section by describing key parts of the Formative Assessment Lesson "Interpreting Distance-Time Graphs." The full lesson can be downloaded from https://www.map.mathshell.org/lessons.php?unit=8225&collection=8.

A few days before the lesson the teacher assigns the ungraded pre-assessment shown in Figure 7.3 as homework.

Shape Statements

1. **James says:**

> If you draw two shapes, the shape with the greater area will also have the longer perimeter.

Is James' statement Always, Sometimes or Never True?

Fully explain and illustrate your answer.

2. **Clara says:**

> If you join the midpoints of the opposite sides of a trapezoid, you split the trapezoid into two equal areas.

Is Clara's statement Always, Sometimes or Never True?

Fully explain and illustrate your answer.

Figure 7.2 Tasks that reveal misunderstandings and provoke rich conversations.

Journey to the Bus Stop

Every morning Tom walks along a straight road from his home to a bus stop, a distance of 160 meters. The graph shows his journey on one particular day.

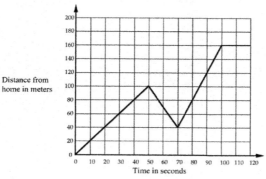

1. Describe what may have happened. You should include details like how fast he walked.
2. Are all sections of the graph realistic? Fully explain your answer.

Figure 7.3 The pre-assessment for the distance-time graphs lesson. 1. Describe what may have happened. You should include details like how fast he walked. 2. Are all sections of the graph realistic? Fully explain your answer.

Typically, some students' papers contain stories that describe Tom walking up and down a hill. While this is wrong (see *x-y* axes), it's important to acknowledge why a student might associate the shape of the graph with walking up and down a hill before shifting towards a mathematically valid interpretation. Even for students who understand that the graph represents how rapidly Tom is walking and whether he's walking to or away from home, interpreting the horizontal segment of the graph at the end of his walk can be a challenge.

The lesson itself begins with the whole-class activity shown in Figure 7.4. Students are given a few minutes to decide, individually, which story matches the graph, and to be ready to provide at least two reasons that justify their choice. Then the teacher asks for a show of hands. Who chose option A, and why? The same is done for options B and C.

Matching a graph to a story

Typically a substantial number of students choose each of the three options. Given these conflicting opinions there's a lot to talk about, so there's a whole class discussion to analyze what each segment of the graph represents. The discussion makes sure that some basics are agreed upon, so the students are ready for the lesson's main activity, a "card sort."

Students work in small groups for the card sort. Each group is provided with a collection of 30 cards that contains 10 stories, 10 graphs, and 10 data tables. Sample stories, graphs, and tables are given in Figure 7.5.

Their next assignment is to match stories, graphs, and tables – to sort through the 30 cards and identify 10 groupings of 3 cards that each represent the same distance-time narrative – and then to put together a poster that presents its groupings. (Figure 7.6 presents part of a poster.) There is then a carefully orchestrated activity in which groups compare and contrast their posters, making revisions if they think they are necessary. This is followed

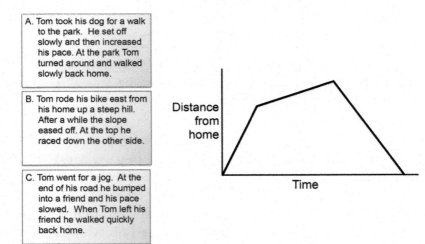

Figure 7.4 The first task in the lesson. Which story matches the graph?

by a whole class discussion of particularly challenging issues and additional activities.

The entire FAL, including the additional activities, takes two or three class periods. The lesson plan, which is 23 pages long (!) helps to prepare the teacher for what to expect as well as to structure the activity. In what follows we'll refer to the lesson plan and to what happens during a typical lesson in order to illustrate the key points given in italics.

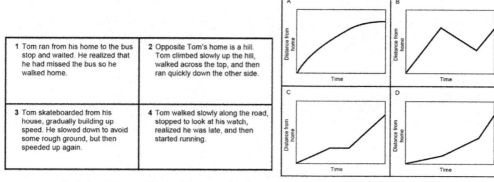

4 of the 10 sample stories   4 of the 10 sample graphs

4 of the 10 sample tables

Figure 7.5 Some of the stories, graphs, and tables in the card sort activity.

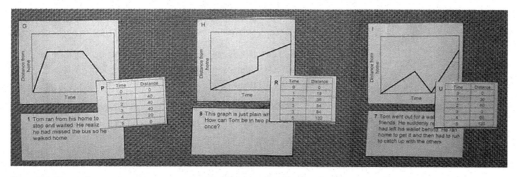

Figure 7.6 Part of a student poster grouping stories, graphs, and tables.

The more you know about student misconceptions, the more you can anticipate them and bring them to light.

A large amount of knowledge about student misconceptions has been systematized. Each of the FALs addresses student misconceptions directly, providing a list of "common issues" in student understanding and a set of questions and prompts aimed at helping students address those issues. The first two issues discussed in the Distance-Time Graphs FAL are given in Figure 7.7.

Once issues such as these have been brought out into the open, the best way to address them is through sense-making.

As we've noted, "fixing" student errors by showing students (again!) how to do things "the right way" tends not to work. The goal is learning with understanding. The more coherently knowledge is tied to a network of

Common issues	Suggested questions and prompts
(Issue 1) Student interprets the graph as a picture. For example: The student assumes that as the graph goes up and down, Tom's path is going up and down. Or: The student assumes that a straight line on a graph means that the motion is along a straight path. Or: The student thinks the negative slope means Tom has taken a detour.	• If a person walked in a circle around their home, what would the graph look like? • If a person walked at a steady speed up and down a hill, directly away from home, what would the graph look like? • In each section of his journey, is Tom's speed steady or is it changing? How do you know? • How can you figure out Tom's speed in each section of the journey?
(Issue 2) Student interprets graph as speed–time The student has interpreted a positive slope as speeding up and a negative slope as slowing down.	• If a person walked for a mile at a steady speed, away from home, then turned round and walked back home at the same steady speed, what would the graph look like? • How does the distance change during the second section of Tom's journey? What does this mean? • How does the distance change during the last section of Tom's journey? What does this mean? • How can you tell if Tom is traveling away from or towards home?

Figure 7.7 Common "issues" and ways to respond without giving the game away.

understandings and constructed by the students themselves, the more robust and resistant to error it is. Figure 7.7 attends to that issue. All of the responses to common student issues consist of *questions* that are intended to help the student make deeper and richer mathematical connections. When the students can answer those kinds of questions, misunderstandings are less likely.

Students are essential resources for instruction.

In classrooms where students are expected to explain their thinking and to listen and respond thoughtfully to each other, a huge amount of teaching and learning takes place in student-to-student conversations.

Consider the card matching task in the Distance-Time Graphs FAL. Teams work together to match stories with graphs. If two students pick different stories to match the same graph or two graphs for the same story, they know there's a problem – and that they need to reason their way through it. It's in situations like this, when students take their responsibility to mathematics and to each other seriously, that a lot of learning takes place! For detailed examples of this, see the case studies in Schoenfeld et al. (2023).

Task design is really important... but equally important is being attuned to student thinking and leveraging it in the moment.

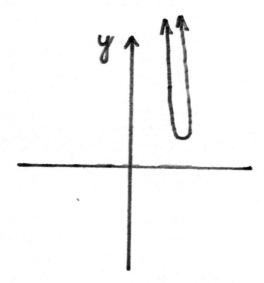

Figure 7.8 Can a very skinny parabola live in just one quadrant? From Schoenfeld et al. in press, with permission.

Well-designed tasks help bring student thinking to the fore, setting up rich conversations around fundamental ideas. Equally important, however, is the act of listening carefully to what our students say! Student conversations will often reveal non-standard ways of thinking – and if we leverage it, that can lead to fascinating mathematical discussions. In a lesson on graphing quadratic functions, for example, the teacher overheard a student in a small group saying, "if a parabola is really skinny, can it live in just one quadrant?" See Figure 7.8.

This one comment led to a very rich classroom discussion. Some of the issues that came up included: Just what is the domain of a quadratic function, and what does that imply about intercepts? Are there circumstances in real-world modeling problems when the function would only be defined for positive x? And so on.

More specific questions for planning and review

Here are some specific questions you might consider as you think about opportunities to implement and expand formative assessment. These are taken from the *TRU Math Conversation Guide* (Baldinger, Louie, and the Algebra Teaching Study and Mathematics Assessment Project, 2018).

- What opportunities exist (or could exist) for students to develop their own strategies, approaches, and understandings of mathematics?
- What opportunities exist (or could exist) for students to share their ideas and reasoning and to connect their ideas to others'?
- What different ways do students get to share their mathematical ideas and reasoning (writing on paper, speaking, writing on the board, creating diagrams, demonstrating with materials/artifacts, etc.)? Who do students get to share their ideas with (a partner, a small group, the whole class, the teacher)?
- What opportunities exist to build on students' mathematical thinking, and how are teachers and/or other students taking up these opportunities?
- How do students seem to be making sense of the mathematics in the lesson, and what responses might build on that thinking?
- How can activities be structured so that students have more opportunity to build on each other's ideas?

– What might we try (what tasks, lesson structures, questioning prompts, etc.) to surface student thinking, especially the thinking of students whose ideas we don't know much about yet?

Some resources that might be useful

Far and away the best resource for formative assessment is the collection of Formative Assessment Lessons, downloadable at no cost from the Mathematics Assessment Project at https://www.map.mathshell.org/lessons.php. There are 20 FALs for each of sixth, seventh, eighth, and ninth grades, and "high school."

Independent research (Herman et al., 2014; Research for Action, 2015) indicated that teaching 4–6 of the FALs over the course of a year resulted in *very* substantial learning gains – on much more than the content covered. Why? In brief, because the lessons scaffolded a change in the ways that the teachers who used the FALs taught their everyday lessons. As they got more comfortable implementing the FALs, teachers wound up doing less "telling" and spent much more time supporting and scaffolding student dialogue in their everyday instruction (Kim, 2017).

Appendix A contains a collection of classroom strategies that serve multiple purposes. Specific strategies that are often useful for formative assessment include *What's the big picture?*, *Share what your partner said*, *Invent an argument*, *Fishbowl*, *Open it up*, *Reflective journals*, *Student-led questioning routine*, *Gallery walk*, and *Group questions.*

In addition to the articles referred to in this section, some papers that go into greater depth are: Black and Wiliam (2009), Kingston and Nash (2011), Kluger and Denisi (1996), Stenmark (1991), Swan (2006), Swan and Burkhardt (2014), Wiliam (2017), Wiliam and Thompson (2007).

The formative assessment targets

This is our final set of targets! Figure 7.9 lists the core questions for the Formative Assessment targets; Figures 7.10–7.12 elaborate on those core questions. You may want to articulate your own goals or ideas on the target in Figure 7.13. Take a look at the tasks you're planning to (or did) use. See if you can place them on the targets and if you can modify them so that they move more toward the center.

Formative Assessment

To what extent is students' mathematical thinking surfaced; to what extent does instruction build on student ideas when potentially valuable or address emerging misunderstandings?

Core Questions:

1. In what ways are student ideas, strategies and reasoning processes brought out into the open?

2. In what ways are students' informal understandings and language use valued and built on?

3. In what ways do whole class activities or group interactions support the refinement of student thinking?

Figure 7.9 Core questions about Dimension 5, Formative Assessment.

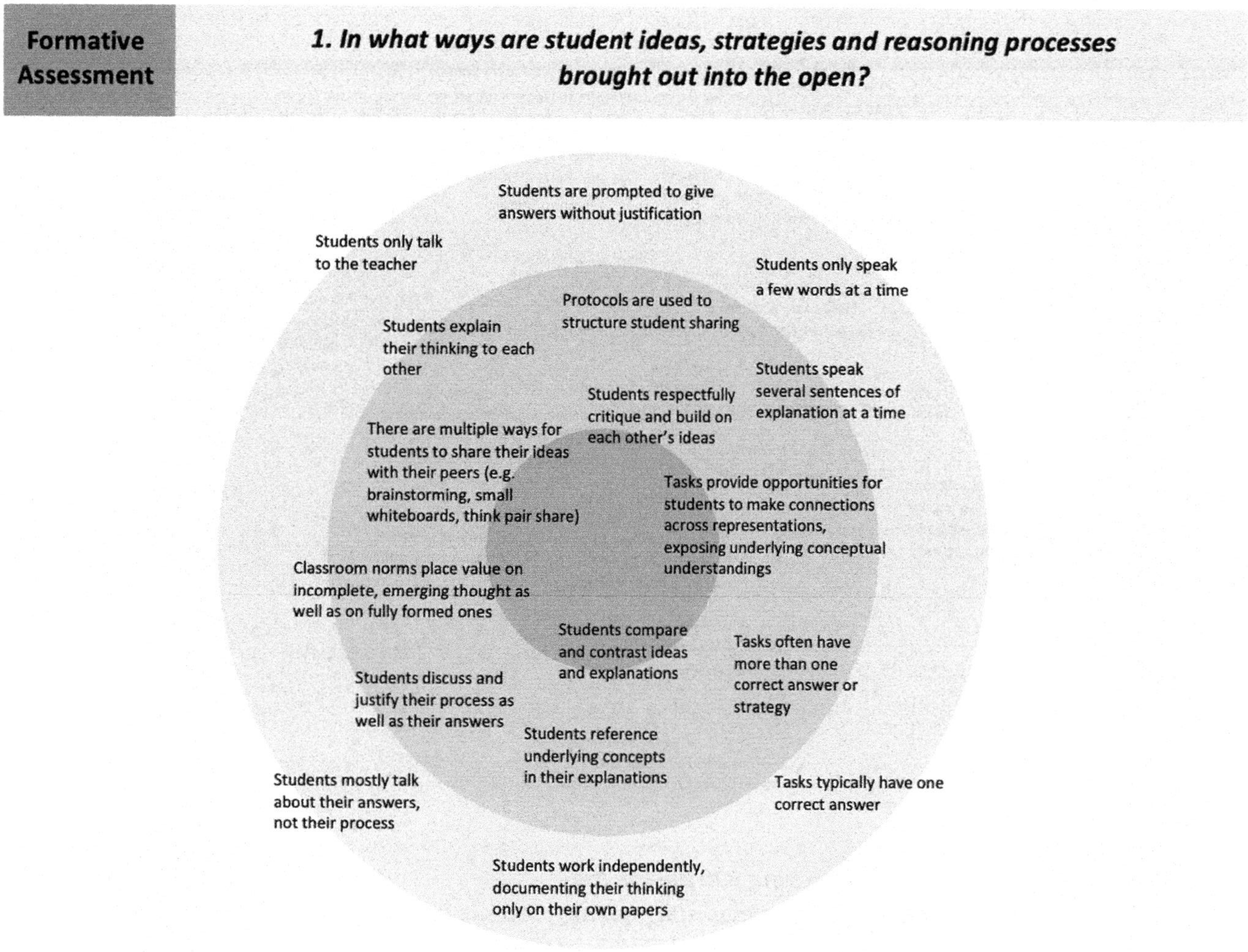

Formative
Assessment

1. In what ways are student ideas, strategies and reasoning processes brought out into the open?

Students only talk
to the teacher

Students are prompted to give
answers without justification

Students only speak
a few words at a time

Protocols are used to
structure student sharing

Students explain
their thinking to each
other

Students speak
several sentences of
explanation at a time

There are multiple ways for
students to share their ideas
with their peers (e.g.
brainstorming, small
whiteboards, think pair share)

Students respectfully
critique and build on
each other's ideas

Tasks provide opportunities for
students to make connections
across representations,
exposing underlying conceptual
understandings

Classroom norms place value on
incomplete, emerging thought as
well as on fully formed ones

Students compare
and contrast ideas
and explanations

Tasks often have
more than one
correct answer or
strategy

Students discuss and
justify their process as
well as their answers

Students reference
underlying concepts
in their explanations

Students mostly talk
about their answers,
not their process

Students work independently,
documenting their thinking
only on their own papers

Tasks typically have one
correct answer

Figure 7.10 Formative Assessment Target 1, bringing student thinking out into the open.

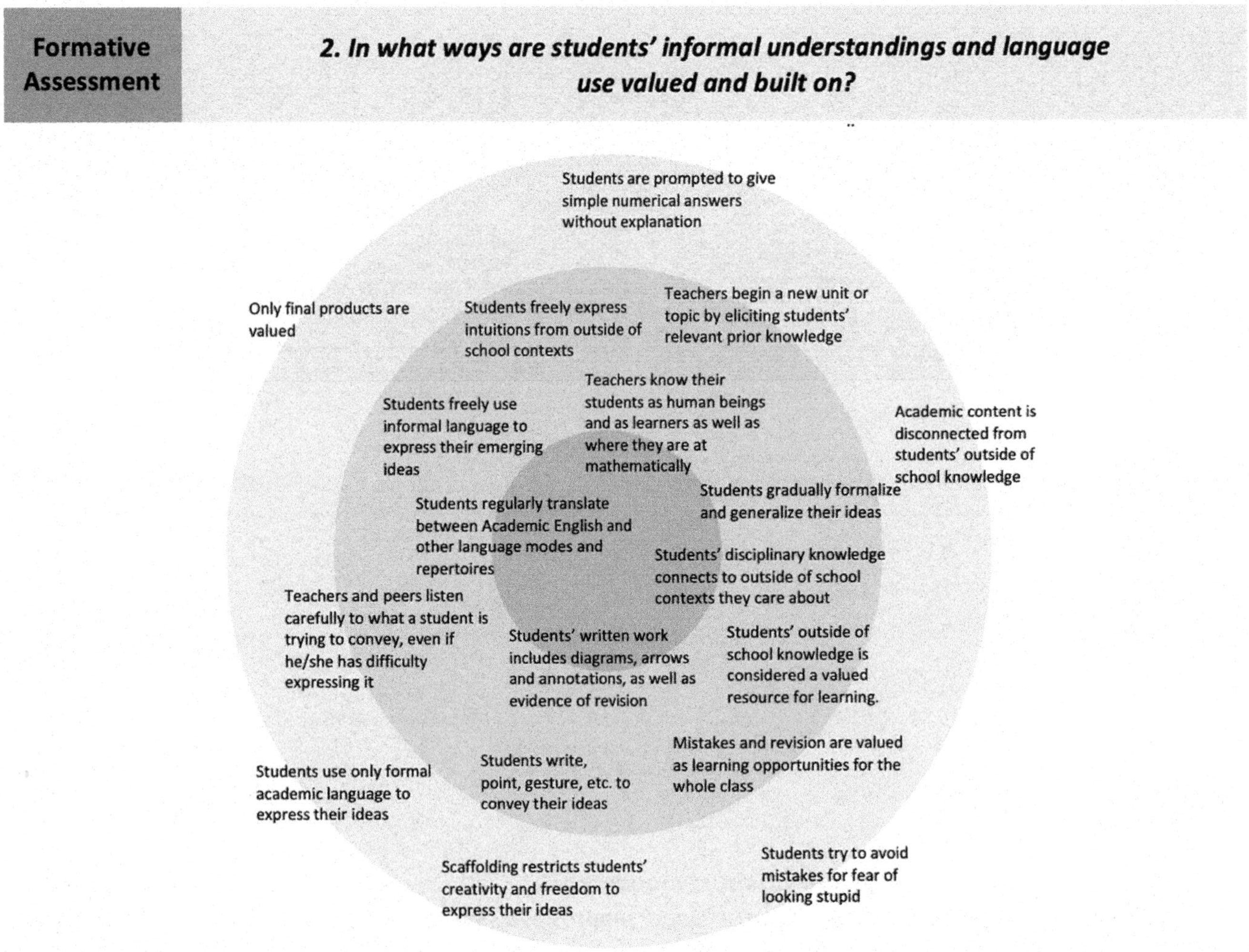

Formative Assessment

2. In what ways are students' informal understandings and language use valued and built on?

Students are prompted to give simple numerical answers without explanation

Teachers begin a new unit or topic by eliciting students' relevant prior knowledge

Students freely express intuitions from outside of school contexts

Teachers know their students as human beings and as learners as well as where they are at mathematically

Students gradually formalize and generalize their ideas

Academic content is disconnected from students' outside of school knowledge

Students freely use informal language to express their emerging ideas

Students' disciplinary knowledge connects to outside of school contexts they care about

Students' outside of school knowledge is considered a valued resource for learning.

Students regularly translate between Academic English and other language modes and repertoires

Students' written work includes diagrams, arrows and annotations, as well as evidence of revision

Mistakes and revision are valued as learning opportunities for the whole class

Students try to avoid mistakes for fear of looking stupid

Only final products are valued

Teachers and peers listen carefully to what a student is trying to convey, even if he/she has difficulty expressing it

Students write, point, gesture, etc. to convey their ideas

Students use only formal academic language to express their ideas

Scaffolding restricts students' creativity and freedom to express their ideas

Figure 7.11 Formative Assessment Target 2, building on students' informal language and understandings.

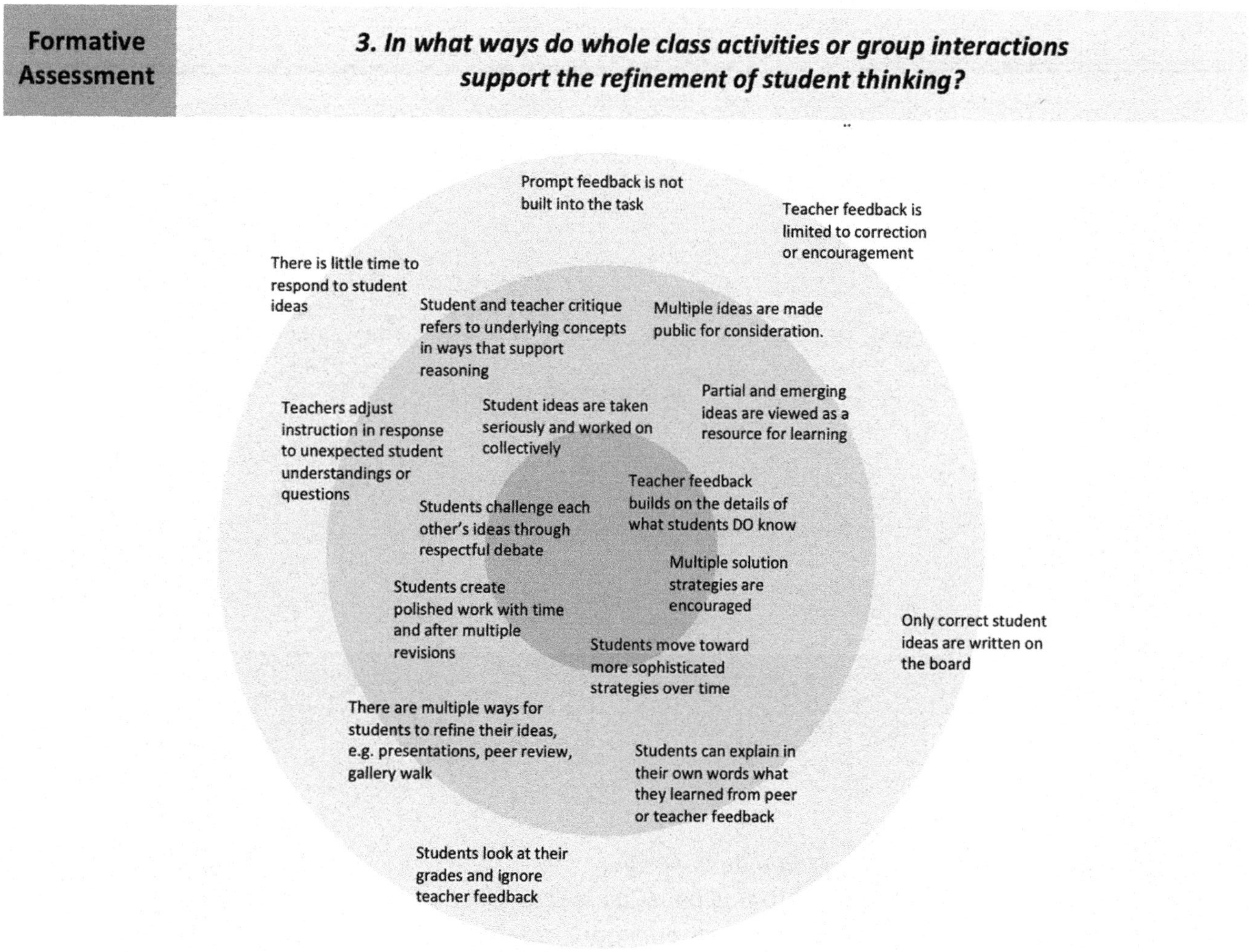

Formative
Assessment

3. In what ways do whole class activities or group interactions support the refinement of student thinking?

Teacher feedback is limited to correction or encouragement

Prompt feedback is not built into the task

Only correct student ideas are written on the board

Multiple ideas are made public for consideration.

Partial and emerging ideas are viewed as a resource for learning

There is little time to respond to student ideas

Student and teacher critique refers to underlying concepts in ways that support reasoning

Student ideas are taken seriously and worked on collectively

Teacher feedback builds on the details of what students DO know

Multiple solution strategies are encouraged

Teachers adjust instruction in response to unexpected student understandings or questions

Students challenge each other's ideas through respectful debate

Students move toward more sophisticated strategies over time

Students can explain in their own words what they learned from peer or teacher feedback

Students create polished work with time and after multiple revisions

There are multiple ways for students to refine their ideas, e.g. presentations, peer review, gallery walk

Students look at their grades and ignore teacher feedback

Figure 7.12 Formative Assessment Target 3, refining student thinking.

Formative Assessment

To what extent is students' mathematical thinking surfaced; to what extent does instruction build on student ideas when potentially valuable or address emerging misunderstandings?

Feel free to add your own!

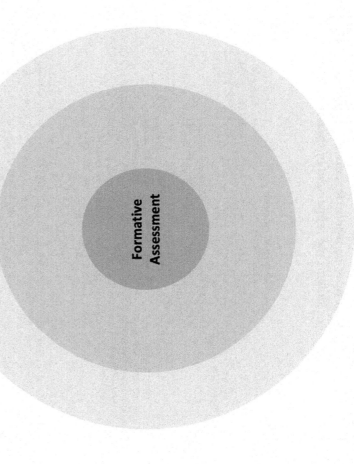

Formative Assessment

Figure 7.13 Make your own formative assessment target!

Using the formative assessment targets – three examples

Here are our Formative Assessment examples, in ascending grade order.

An elementary grades example – the mosaic fraction task

Task enhancement strategy: encourage varied forms of justification

Fractions are a major focus of instruction in elementary classrooms, specifically in grades 3 through 5. If you google "fractions worksheets" and click on google images, you'll find hundreds of worksheets, almost all of which focus on one aspect of fractions, such as equivalent fractions, part-whole models, and size comparisons. Almost all of them raise the same issues. See Figure 7.14 for a screen shot of the search page. What can we do in the face of such repetitive exercises?

Identifying the challenges

On Target 1:

- Students work independently, only documenting their thinking on their papers.
- Tasks typically have one correct answer.

On Target 2:

- Scaffolding restricts students' creativity and freedom to express their ideas.

On Target 3:

- Teacher feedback is limited to correction or encouragement.

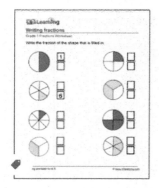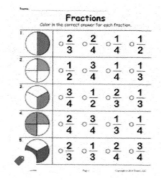

Figure 7.14 Common fraction practice assignments.

Identifying the opportunities

On Target 1:

- Students discuss and justify their process as well as their answers.

On Target 2:

- Students gradually formalize and generalize their ideas.

On Target 3:

- Multiple solution strategies are encouraged.

Generating possibilities

The narrow framing in the tasks shown in Figure 7.15 removes challenge and leaves little room for students to think for themselves. It offers little if any opportunity for the teacher or other students to understand how specific students are making sense of fractional comparisons. Students could complete this activity very quickly, with answers being the presumed goal as opposed to deeper conceptual understanding.

Fractional comparisons are a key component of elementary mathematics, worthy of extended discussions and investigations. Students need opportunities to combine their intuitions with mathematical reasoning and justifications. The task presented in Figure 7.15 could be given in the middle or the end of a unit, as a formative and/or summative assessment.[1] It could also be

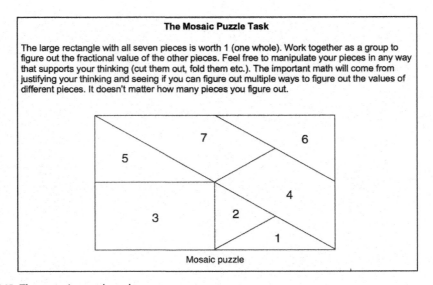

The Mosaic Puzzle Task

The large rectangle with all seven pieces is worth 1 (one whole). Work together as a group to figure out the fractional value of the other pieces. Feel free to manipulate your pieces in any way that supports your thinking (cut them out, fold them etc.). The important math will come from justifying your thinking and seeing if you can figure out multiple ways to figure out the values of different pieces. It doesn't matter how many pieces you figure out.

Mosaic puzzle

Figure 7.15 The mosaic puzzle task.

presented at the beginning of a unit on fractional comparisons, with students working on the task part by part over the span of several days or weeks as they learn related skills in class.

Regardless of when this task is presented to students, the intent is to prompt students to think creatively, to talk to each other, and to draw on various fraction-related understandings to communicate their ideas and justify their conclusions. Note that the multiple ways to enter into the problem and to engage in conversation about it not only has the potential to engage the students with rich mathematics but provides opportunities for access in various ways, and for dialogue that – if the right norms and support structures are in place – can support the development of AOI.

A middle grades example – interpreting graphs

Task enhancement strategy: incorporate card sorts

In middle school, students explore functional relationships between two quantities. Classroom activities often include asking students to make connections between graphs and stories. Exercises involving distance-time graphs are common. Figure 7.16 shows a typical example.

Identifying the challenges

On Target 1:

–   Students work independently, documenting their thinking only on their papers.

On Target 2:

–   Students are prompted to give simple numerical answers without explanation.

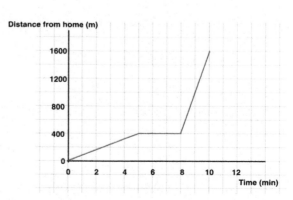

John went to meet a friend for coffee. The graph to the right shows how far John was from home as he walked to get there.

• How far away is the coffee shop?

• Which segment shows him going fastest?

• How fast was he traveling for the first 5 minutes?

• How long was he standing still?

Figure 7.16 A typical distance-time graphing assignment.

On Target 3:

– Prompt feedback is not built into the task.

Identifying the opportunities

On Target 1:

– Students reference underlying concepts in their explanations.

On Target 2:

– Students write, point, gesture, etc., to convey their ideas.

On Target 3:

– Students challenge each other's ideas through respectful debate.

Generating possibilities

This "real-world" context in Figure 7.16 *might* support opportunities for mathematical reasoning. However, the problem is written as an independent task in which students are prompted to give numerical answers only, without explanation or justification. If a student wrote an incorrect answer on their paper, they would likely not realize it until their teacher pointed it out. And if the teacher reviewed student papers without the students being present, the teacher would likely not know how or why the student arrived at their incorrect answer. Even if students got all answers correct, it is still very possible that students' understandings of these concepts were incomplete and/or there was more for them to learn.

Adapting a task into a card-sorting activity, as in the Formative Assessment Lesson discussed in this chapter[2], can support deeper reasoning and help students fine-tune their developing mathematical ideas.

A secondary grades example – ladder problems

Task enhancement strategy: ask open-ended questions before formal instruction

In beginning calculus courses students often encounter "application problems" once they have learned the basic mechanics of differentiation. Students are usually shown how to do a sample problem and then given a series of similar problems on which they practice the solution method. "The Ladder Problem" in Figure 7.17 is a common related rates application problem. A simple Google search will produce a number of different variations, such as the following.

The Ladder Problem (typical version)

A 10-foot ladder leans against a wall. It begins to slide down the wall. Suppose the bottom slides away from the wall at a rate of 1 ft/sec. How fast is the top of the ladder sliding down the wall when the bottom of the ladder is 6 feet from the wall?

Figure 7.17 A Ladder Problem.

Identifying the challenges

On Target 1:

– Students work independently, documenting their thinking only on their papers.

On Target 2:

– Students use only formal academic language to express their ideas.

On Target 3:

– Prompt feedback is not built into the task.

Identifying the opportunities

On Target 1:

– Students compare and contrast ideas and explanations.

On Target 2:

– Students gradually formalize and generalize their ideas.

On Target 3:

– Student ideas are taken seriously and worked on collectively.

Generating possibilities

"Application" problems hold great potential to support students in rich mathematical engagement. However, if math problems are presented in ways that elicit rote responses that mimic what they've been shown in sample problems, then this potential can go unrealized. The Ladder Problem, as presented above, would typically be given as an independent practice problem. After reviewing what the student wrote on their paper, their teacher would have a sense of whether or not this student had applied the intended differentiation procedures correctly. But the teacher likely wouldn't know how the student was thinking about related rates as a concept or if the student knew why they needed to differentiate in this particular situation. If the student got an incorrect answer, the teacher could point to the procedural step at which they went wrong but wouldn't know how to build on what the student was indeed understanding.

Slight tweaks to The Ladder Problem (or other application problems) can improve students' engagement with the mathematics and visibility of student thinking. Consider this enhanced version of The Ladder Problem[3] in Figure 7.18.

This task was given to students as the first problem in the unit on related rates, and students were asked to work on it in small groups during class time. The students had learned various methods of differentiation but had not been shown how to apply those methods to "real-life" situations. The open-ended questions in part 1 prompt students to share intuitions and explanations that may or may not include formalized mathematical thinking. But it's not simply that these questions are open-ended; it's that the questions get the students to consider the underlying situational model. The questions help students understand the context and what the situation is like before jumping into equations. Whether in word problems or calculus or elsewhere, there's a tendency to leap to symbolic representations. But the effective use of mathematics symbols depends on their corresponding to the situation being symbolized. Asking about how the variables in the problem are related helps students develop intuitions that can then serve as "reality checks" on their symbolic manipulations. In addition, giving this problem at the beginning of the unit enabled the teacher to learn how students were making sense of the underlying conceptual understandings that she knew would be critical as they moved forward with the calculus course.

Our three examples have shown ways in which tasks can be modified to stimulate student thinking, bring it out into the open, and have it serve as a resource for classroom conversations. As we noted in the introduction to this dimension, that sets the stage for adapting instruction to "meet the students where they are." We don't have the space to work through examples of

---

The Ladder Problem (enhanced version)

1. A ladder leans against a wall. It begins to slide down the wall. Does the top of the ladder move at the same rate as the bottom of the ladder? How do you know?
2. Suppose the bottom slides away from the wall at a rate of 1 foot per second. How fast is the top of the ladder sliding down the wall when the bottom of the ladder is 6 feet from the wall? Assume the ladder is 10 feet long.

Figure 7.18 The Ladder Problem, enhanced.

teachers and students making real-time, productive adjustments to advance learning – but, please see the three extended classroom studies in our book, *Helping Students Become Powerful Mathematics Thinkers: Case Studies of Teaching for Robust Understanding* (Schoenfeld et al., 2023). We hope both books will support you in your lifelong efforts to create increasingly ambitious and equitable learning environments for your students.

Notes

1   This task was created by Mallika Scott, based on an activity in Van De Walle's, Pearson, and Bay-Williams (2018).
2   See https://www.map.mathshell.org/lessons.php?unit=8225&collection=8. The Formative Assessment Lessons contain a number of card sorting activities.
3   The enhanced version of The Ladder Problem is based on one that was designed and implemented by Masha Albrecht at Berkeley High School. Masha credits Diane Resek, Professor of Mathematics Education at San Francisco State University, for inspiring her design of this task.

8

Appendix
Some useful strategies

1 **What's the big picture?** Identify an "essential question" or "big idea" that's important for the math your students are working on. Share it with students at the beginning of the lesson and find time, especially at the end of the lesson, for students to connect what they're doing to the big idea. Check in with students about it while they're working in groups, or have a few students share or lead a class discussion about it at the end of a lesson.

2 **Share what your partner said** Before a group discussion, have students turn and talk to a neighbor. When the discussion begins, have students who want to talk share their neighbor's idea, not their own.

3 **Ask for patterns and outliers** To give students something more to think about while they practice similar problems or do problem sets, ask them to describe any mathematical patterns they noticed while working on similar problems. Or, ask them to find problems that seemed like outliers – ones for which the method didn't work as expected, or with a solution that surprised or confused them. This can be given in addition to a regular homework assignment. Have a few students share their patterns and outliers in class, or have students discuss them in pairs.

4 **Deep dive homework** Instead of turning in an entire problem set from homework, students pick some small, reasonable number of problems that they will show work from in detail and then turn in.

DOI: 10.4324/9781003376903-8

The goal is to have them make something they want to show the teacher, whether because they're impressed with what they did, want feedback on that specifically, enjoyed solving it, and want to spend more time.

5 **Name It** When a student shares an idea, write their name on the board next to it when you record it. Refer to that idea as "so-and-so's idea" or "so-and-so's strategy." Keep track of whose ideas have been highlighted from class to class.

6 **Invent an argument** Come up with some fictional characters who disagree about a math idea you want to teach and ask students how they would resolve the argument. Make sure students also discuss why each side thinks what they do, even if the characters' arguments aren't correct.

7 **Fishbowl** During class discussions, split the group in half. Have one half sit together so they can see each other and the other half sit in a circle around them. The inner group talks for a certain amount of time (3 minutes, 5 minutes) with minimal input from you. The outer group writes what they heard during the discussion. Collect their notes; switch groups on different days.

8 **Make categories** After a long problem set or homework assignment filled with similar problems, ask students to group the problems into categories. They can choose the categories they like but encourage them to be mathematical. For example, after a homework assignment of solving systems of equations, students might choose to categorize the problems according to those that have solutions and those that do not; those that have solutions in different quadrants when graphed; those that were best solved with different methods; etc. Let students be creative! This gives students opportunities to see the bigger picture after doing detailed problems.

9 **Open it up** Reframe a problem or problem set so that there's more for students to discover. For example, when doing proofs with students, instead of asking them to "show" something, such as that the diagonals of a rectangle bisect each other (which assumes that it's already known to be true), pose the thing to be shown as something to discover. Ask, "What do you notice about the diagonals of different quadrilaterals?" and give students time to draw, find patterns, and make claims about what they see.

10 **Wait a turn** During whole class discussions, try letting two (three, four, five…) students speak in a row before you say something.

11 **Reflective journals** can begin simply, by asking students to respond to a single question like: What did I learn today? Every day, teachers can choose three journal entries to read to everyone at the beginning of the next lesson that revisit content, embody habits they are trying to build in the whole class, etc. Over a couple of weeks, all students' journals are chosen. Journals can be a place where students write the problem, record their thinking: how it changed and why, and reflect on what they did not know previously that they know now.

12 **Confidence thermometer** At the end of a lesson or unit, ask students to rate how comfortable they are with key idea(s), by coloring in an empty bar (or thermometer) next to each key idea. Students who feel comfortable with a given idea can share tips for studying it or making sense of it.

13 **Time alone** Give students some time (maybe 5 minutes) to work alone on a problem before working in a group, so that they have time to think and prepare their ideas. You may want to structure this "alone time" by setting expectations as to what students will share with their groups after 5 minutes. For example, tell students, "Be prepared to share one idea that you have for getting started on this problem and one question that you have about the problem." or "Be prepared to share one thing that you know about this problem and one thing that you don't know about this problem."

14 **Student-led questioning routine** Establish a routine of student-led questions and challenges after viewing each student presentation or work. For example, students might routinely finish their presentation by asking the class "Do you have comments or questions?" and calling on classmates.

15 **What do you wish you knew?** At the end of a test or quiz, ask students to choose at least one problem they aren't sure they did correctly and write a quick reflection about what they think they're missing or what tool/method/idea would be useful to solve the problem correctly.

16 **Share wrong answers like they're right answers** Share an incorrect solution to a problem, saying "another student solved the problem this way. Identify similarities and differences between this solution and the solution you have. What makes sense to you? What questions do you have?" Encourage students to share their thinking in a whole class discussion or in small groups about their own solution to the problem as it relates to the incorrect solution the teacher provides. If possible, let the students come to the decision that the provided solution is incorrect and encourage them to justify their thinking.

17 **Three things** If students are stuck on a problem, ask them to state three things they know about the problem and three things they are wondering about. This can help students realize that they are making progress and have information that they can potentially draw on to make more progress.

18 **Choose your color** Each student uses a different color pen/pencil when working on a group problem. Students can use poster paper or larger than normal paper, so that everyone has physical access to the paper.

19 **Extended wait time** After posing a question to the whole class, count to five slowly in your head before calling on a student, hopefully resulting in more hands being raised.

20 **Gallery walk** Students post their individual or group work around the room and then walk around and look at their peers' work. Work may be in finished form or in progress. Gallery walks can be "silent" or interactive. Students can post comments/question using sticky notes or talk through questions and comments. With group work, one student can be chosen to stay behind to present their group's work to the other students.

21 **Always, Sometimes, Never** Pose questions that require explanation and are not straight yes/no or one answer solutions. For example, instead of asking, "Is it true or false that when you multiply two numbers, the answer will always be bigger?" ask, "When you multiply two numbers, the answer will be bigger. Is this statement always, sometimes or never true? Explain your reasoning." This gives students opportunities to explore more possibilities, such as positive/negative numbers and fractions.

22 **Group questions** The teacher only answers group questions (questions that have been agreed upon by all group members). One student raises their hand when the group has a question. The teacher asks a different student in the group what the question is and only answers the question if all group members agree it's a question.

23 **Idea melting-pot** After a do-now, rather than having students raise their hands to share how they solved a problem, have them turn in their work to you without their names on it. Randomly pick one of the papers and choose a student to present the solution written there. The class can work together to understand what was done.

24 **Start with the end** Start a unit with a rich problem that draws out ideas that will be useful throughout the unit and helps students build them. In some textbooks, these problems typically occur at the end of the unit. In contrast, some research-based textbooks use rich

problems to *begin* the unit, so that students can actually develop new procedures or ideas of the unit using their prior knowledge. Such experiences help students realize they can develop procedures that they have not yet been taught.

25 **The three reads protocol** This strategy involves having students read a math scenario three times, with a different goal each time. The first read is to understand the context. The second read is to understand the mathematics. The third read is to elicit inquiry questions based on the scenario. The intent is to support students in learning to do a close read of complex math word problems or tasks. (This is one of many powerful language and reading-related strategies. See Zwiers et al., 2017.)

References

American Association of University Women. (1992). *How Schools Shortchange Girls*. Annapolis Junction, MD: AAUW.

Baldinger, E., Louie, N., and the Algebra Teaching Study and Mathematics Assessment Project. (2018). *TRU Math conversation guide: A tool for teacher learning and growth (mathematics version)*. Berkeley, CA & E. Lansing, MI: Graduate School of Education, University of California, Berkeley & College of Education, Michigan State University. Retrieved from: https://truframework.org.

Banks, J., & Banks, C. (Eds.). (2004). *Handbook of research on multicultural education* (2nd edition). New York: Jossey-Bass.

Bartell, T., Berry, R., Felton-Koestler, M., & Yeh, C. (2020). *Upper elementary mathematics lessons to explore, understand, and respond to social injustice.* Thousand Oaks, CA and Reston, VA: Corwin Press and NCTM.

Berry, R., Conway, B., Lawler, B., & Staley, J. (2020). *High school mathematics lessons to explore, understand, and respond to social injustice.* Thousand Oaks, CA and Reston, VA: Corwin Press and NCTM.

Black, P. J., & Wiliam, D. (1998). Assessment and classroom learning. *Assessment in Education*, 5(1), 7–74.

Black, P. J., & Wiliam, D. (2009). Developing the theory of formative assessment. *Educational Assessment, Evaluation and Accountability*, 21(1), 5–31.

Boaler, J. (2008). Promoting relational equity in mathematics classrooms – Important teaching practices and their impact on student learning. *Proceedings of the 10th International Congress of Mathematics Education (ICME X)*, 2004, Copenhagen.

Boaler, J., & Staples, M. (2008). Creating mathematical futures through an equitable teaching approach: The case of Railside School. *The Teachers College Record*, 110(3), 608–645.

Brown, A. L., Ash, D., Rutherford, M., Nakagawa, K., Gordon, A., & Campione, J. C. (1993). Open it up, Wait a turn. Distributed expertise in the classroom. In G. Salomon (Ed.), *Distributed cognitions: Psychological and educational considerations* (pp. 188–228). New York: Cambridge University Press.

Burkhardt, H., & Schoenfeld, A. H. (2019). Formative assessment in mathematics. In R. Bennett, H. Andrade, & G. Cizek (Eds.), *Handbook of Formative Assessment in the Disciplines* (pp. 35–67). New York: Routledge. ISBN 9781138054363.

Burkhardt, H., & Schoenfeld, A. H. (2022). Assessment and mathematical literacy: A brief introduction. In R.J. Tierney, F. Rizvi, F., & K. Erkican (Eds.), *International Encyclopedia of Education*, 4th Edition, vol. 13. Elsevier. https://dx.doi.org/10.1016/B978-0-12-818630-5.09007-2. ISBN: 9780128186305.

Burns, M. (2004). Writing in math. *Educational Leadership, 62*(2), 30–33.

Cohen, E. G., & Lotan, R. A. (Eds.). (1997). *Working for equity in heterogeneous classrooms: Sociological theory in practice.* New York: Teachers College Press.

Cohen, E. G., & Lotan, R. A. (2014). *Designing groupwork: Strategies for the heterogeneous classroom* (Third Edition). New York: Teachers College Press.

Cohen, E. G., Lotan R. A., Scarloss, B. A., & Arellano, A. R. (1999). Complex instruction: Equity in cooperative learning classrooms, *Theory into Practice, 38*(2), 80–86. https://doi.org/10.1080/00405849909543836.

Common Core State Standards Initiative (2010). *Common core state standards for mathematics.* Downloaded June 4, 2010 from http://www.corestandards.org/the-standards.

Conway, B., Id-Deen, L., Raygoza, M., Ruiz, A., Staley, J., & Thanheiser, E. (in press). *Middle school mathematics lessons to explore, understand, and respond to social injustice.* Thousand Oaks, CA and Reston, VA: Corwin Press and NCTM.

Cuoco, A., Goldenberg, E. P., & Mark, J. (1996). Habits of mind: An organizing principle for mathematics curricula. *Journal of Mathematical Behavior, 15*, 375–402. Downloaded September 15, 2022, from https://nrich.maths.org/content/id/9968/Cuoco_etal-1996.pdf.

Darling Hammond, L. (2010). *The flat world and education: How America's commitment to equity will determine our future.* New York: Teachers College Press.

Daro, V. (2021). Growing student language in math class. https://envisionlearning.org/wp-content/uploads/2022/01/ELP-Growing-Student-Language-in-Math-Class-2021.pdf.

Diversity in Mathematics Education (DiME) Center for learning and teaching. (2007). Culture, race, power, and mathematics education. In F. Lester (Ed.), *Handbook of research on mathematics teaching and learning* (Second Edition, pp. 405–434). Charlotte, NC: Information Age Publishing.

Dweck, C. (2007). *Mindset: The new psychology of success.* New York: Ballantine.

Dweck, C. S. (2012). *Mindset: How you can fulfill your potential.* Constable & Robinson Limited.

Engle, R. A. (2011). The productive disciplinary engagement framework: Origins, key concepts, and continuing developments. In D. Y. Dai (Ed.), *Design research on learning and thinking in educational settings: Enhancing intellectual growth and functioning* (pp. 161–200). London: Taylor & Francis.

Esmonde, I. (2009). Ideas and identities: Supporting equity in cooperative mathematics learning. *Review of Educational Research, 79*(2), 1008–1043.

Fink, H. (2022). Centering students' voices: A multifocal mixed methods investigation of participatory equity in a distance learning calculus class. Ph.D. Dissertation, University of California, Berkeley.

Gresalfi, M., Martin, T., Hand, V., & Greeno, J. (2009). Constructing competence: An analysis of student participation in the activity systems of mathematics classrooms. *Educational Studies in Mathematics, 70*(1), 49–70.

Gutierrez, K., & Rogoff, B. (2003). Cultural ways of learning: Individual traits or repertoires of practice. *Educational Researcher, 32*(5), 19–25.

Gutstein, R., & Peterson, B. (2013). *Rethinking mathematics: Teaching social justice by the numbers* (second edition). Milwaukee, WI: Rethinking Schools.

Henningsen, M., & Stein, M. K. (1997). Mathematical tasks and student cognition: Classroom-based factors that support and inhibit high-level mathematical thinking and reasoning. *JRME, 28*(5), 524–549.

Herman, J., Epstein, S., Leon, S., La Torre Matrundola, D., Reber, S., & Choi, K. (2014). *Implementation and effects of LDC and MDC in Kentucky districts (CRESST Policy Brief No. 13)*. Los Angeles: University of California, National Center for Research on Evaluation, Standards, and Student Testing (CRESST).

Horn, I. S. (2007). Fast kids, slow kids, lazy kids: Framing the mismatch problem in math teachers' conversations. *Journal of the Learning Sciences, 16*(1), 37–79.

Horn, I. S. (2012). *Strength in numbers: Collaborative learning in secondary mathematics*. Reston, VA: NCTM.

Humphreys, C., & Parker, R. (2015). *Making number talks matter*. Portland, ME: Stenhouse.

Kim, H-j. (2017). Teacher learning opportunities provided by implementing formative assessment lessons: Becoming responsive to student mathematical thinking. *International Journal of Science and Mathematics Education*. Downloaded December 1, 2017, from https://doi.org/10.1007/s10763-017-9866-7.

Kingston, N., & Nash, B. (2011). Formative assessment: A meta-analysis and a call for research. *Educational Measurement: Theory & Practice, 30*(4), 28–37.

Kluger, A. N., & Denisi, A. (1996) The effects of feedback interventions on performance: A historical review, a meta-analysis, and a preliminary feedback intervention theory. *Psychological Bulletin, 119*, 254–284.

Kozol, J. (1992). *Savage inequalities*. New York: Harper Perennial.

Ladson-Billings, G. (1994). *The dreamkeepers: Successful teachers of African American children*. San Francisco: Jossey-Bass.

Ladson-Billings, G. (1995). Toward a theory of culturally relevant pedagogy. *American Research Journal, 32*(3), 465–491.

Ladson-Billings, G. J., & Tate, W. (2006). *Education research in the public interest: Social justice, action, and policy*. New York: Teachers College Press.

Lappan, G., & Phillips, E. (2009). Challenges in U.S. mathematics education through a curriculum developer lens. *Educational Designer*, 1(3). Downloaded January 10, 2009, from http://www.educationaldesigner. org/ed/volume1/issue3/article11/.

Lindberg, S. M., Hyde, J. S., Petersen, J. L., & Linn, M. C. (2010). New trends in gender and mathematics performance: A meta-analysis. *Psychological Bulletin*, 136(6), 1123–1135.

Lotan, R. (2003, March). Group worthy tasks. *Educational Leadership*, 60(6), 72–75.

Martin, D. B. (Ed.) (2009). *Mathematics teaching, learning, and liberation in the lives of black children.* New York: Routledge.

Mason, J., Burton, L., & Stacey, K. (2010). *Thinking mathematically* (Second edition). New York: Prentice-Hall.

Mathematics Assessment Project web site: https://www.map.mathshell.org/.

Mathematics Assessment Project. (2022). *Evaluating statements about length and area*, at http://map.mathshell.org/download.php?fileid=1750.

Maurer, S. (1987). New knowledge about errors and new views about learners: What they mean to educators and what educators would like to know. In A. Schoenfeld (Ed.), *Cognitive science and mathematics education* (pp. 165–187). Mahwah, NJ: Erlbaum.

Michaels, S., O'Connor, M. C., Hall, M. W., & Resnick, L. B. (2010). *Accountable talk sourcebook.* Pittsburgh, PA: Institute for Learning.

Moll, L., Amanti, C., Neff, D., & Gonzalez, N. (1992). Funds of knowledge for teaching: Using a qualitative approach to connect homes to classrooms. *Theory into Practice*, 31(2), 132–141.

Moschkovich, J. N. (2012). *Mathematics, the Common Core, and language: Recommendations for mathematics instruction for ELs aligned with the Common Core.* Proceedings of the "Understanding Language" Conference. Stanford, CA: Stanford University. Available online at http://ell.stanford.edu.

Moschkovich, J. N. (2013). Principles and guidelines for equitable mathematics teaching practices and materials for English language learners. *Journal of Urban Mathematics Education*, 6(1), 45–57.

Moses, R. P. (2001). *Radical equations: Math literacy and civil rights.* Boston, MA: Beacon Press.

Nasir, N., Cabana, C., Shreve, B., Woodbury, E., & Louie, N. (Eds). (2014). *Mathematics for equity: A framework for successful practice.* New York: Teachers College Press.

Nasir, N., & Cobb, P. (eds.) (2007). *Improving access to mathematics: Diversity and equity in the classroom.* New York: Teachers College Press.

Nasir, N., & Shah. N. (2011). On defense: African American males making sense of radicalized narratives in mathematics education. *Journal of African American Males in education*, 2(1), 24–45.

National Council of Teachers of Mathematics. (1980). *An agenda for action.* Reston, VA: NCTM. Downloadable from http://www.nctm.org/flipbooks/standards/agendaforaction/index.html.

National Council of Teachers of Mathematics. (1989). *Curriculum and evaluation standards for school mathematics.* Reston, VA: NCTM.

National Council of Teachers of Mathematics. (2000). *Principles and standards for school mathematics.* Reston, VA: NCTM.

National Council of Teachers of Mathematics. (2023). Teaching and learning mathematics with the common core. http://www.nctm.org/Standards-and-Positions/Common-Core-State-Standards/Teaching-and-Learning-Mathematics-with-the-Common-Core/.

National Research Council. (2001). *Adding it up: Helping children learn mathematics.* In J. Kilpatrick, J. Swafford, and B. Findell (Eds.). *Mathematics learning study committee, center for education, division of behavioral and social sciences and education.* Washington, DC: National Academy Press.

Oakes, J. (2005). *Keeping track: How schools structure inequality* (second edition). New Haven, CT: Yale University Press.

Oakes, J., Joseph, R., & Muir, K. (2004). Access and achievement in mathematics and science. In J. A. Banks & C. A. McGee Banks (Eds.), *Handbook of research on multicultural education* (pp. 69–90). San Francisco: Jossey-Bass.

Parrish, S. (2010). *Number talks.* Sausalito, CA: Math Solutions.

Reinholz, D., & Shah, N. (2018). Equity analytics: A methodological approach for quantifying participation patterns in mathematics classroom discourse. *JRME,* 49(2), 140–177.

Research for Action (2015). *MDC's influence on teaching and learning.* Philadelphia, PA: Author. Downloaded March 1, 2015, from https://www.researchforaction.org/publications/mdcs-influence-on-teaching-and-learning/.

Rehmeyer, J. (2014). *Transitions. Critical issues in mathematics education workshop series: Teaching and learning algebra, workshop 5.* Berkeley, CA: Mathematical Sciences Research Institute.

Rosebery, A., Warren, B., & Conant, F. (1992). Appropriating scientific discourse: Findings from language minority classrooms. *Journal of the Learning Sciences,* 2, 61–94.

Schoenfeld, A. H. (1985). *Mathematical problem solving.* Orlando, FL: Academic Press.

Schoenfeld, A. H. (1992). Learning to think mathematically: Problem solving, metacognition, and sense-making in mathematics. In D. Grouws (Ed.), *Handbook for research on mathematics teaching and learning* (pp. 334–370). New York: MacMillan.

Schoenfeld, A. H. (2003). Making mathematics work for all children: Issues of standards, testing, and equity. *Educational Researcher*, 31(1), 13–25.

Schoenfeld, A. H. (2013). Classroom observations in theory and practice. *ZDM, the International Journal of Mathematics Education*, 45, 607–621. https://doi.org/10.1007/s11858-012-0483-1.

Schoenfeld, A. H. (2014, November). What makes for powerful classrooms, and how can we support teachers in creating them? *Educational Researcher*, 43(8), 404–412. https://doi.org/10.3102/0013189X1455.

Schoenfeld, A. H. (2015). Thoughts on scale. *ZDM, the International Journal of Mathematics Education*, 47, 161–169. https://doi.org/10.1007/s11858-014-0662-3.

Schoenfeld, A. H. (2018). Video analyses for research and professional development: The teaching for robust understanding (TRU) framework. In C. Y. Charalambous & A.-K. Praetorius (Eds.), *Studying instructional quality in mathematics through different lenses: In search of common ground. An issue of ZDM: Mathematics Education*. Manuscript available at https://doi.org/10.1007/s11858-017-0908-y.

Schoenfeld, A. (2022). Why are learning and teaching mathematics so difficult? In M. Danesi, (Ed.), *Handbook of cognitive mathematics*. New York: Springer Nature. https://doi.org/10.1007/978-3-030-44982-7_10-1

Schoenfeld, A. H., & Brown, K. (In press). *Civic discourse in secondary mathematics classrooms*. Washington, DC & Reston, VA: National Academy of Education and NCTM.

Schoenfeld, A. H., Baldinger, E., Disston, J., Donovan, S., Dosalmas, A., Driskill, M., Fink, H., Foster, D., Haumersen, R., Lewis, C., Louie, N., Mertens, A., Murray, E., Narasimhan, L., Ortega, C., Reed, M., Ruiz, S., Sayavedra, A., Sola, T., Tran, K., Weltman, A., Wilson, D., & Zarkh, A. (2019). Learning with and from TRU: Teacher educators and the teaching for robust understanding framework. In K. Beswick (Ed.), *International handbook of mathematics teacher education, Volume 4, The mathematics teacher educator as a developing professional*. Rotterdam: Sense Publishers.

Schoenfeld, A. H., Fink, H., & Zuñiga-Ruiz, S., with S. Huang, X. Wei, and B. Chirinda (2023). *Helping students become powerful mathematics thinkers: Case studies of teaching for robust understanding*. New York: Routledge.

Schoenfeld, A. H., and the Teaching for Robust Understanding Project. (2018). *The teaching for robust understanding (TRU) observation guide for mathematics: A tool for teachers, coaches, administrators, and professional learning communities*. Berkeley: Graduate School of Education, University of California, Berkeley. Retrieved from https://truframework.org.

Shah, N. (2017). Race, ideology, and academic ability: A relational analysis of racial narratives in mathematics. *Teachers College Record*, 119(7), 1–42.

Steele, C. (1997). A threat in the air: How stereotypes shape intellectual identity and performance. *American Psychologist, 52*(6), 613–629.

Steele, C., & Aronson, J. (November 1995). Stereotype threat and the intellectual test performance of African Americans. *Journal of Personality and Social Psychology, 69*(5), 797–811.

Stein, M. K., Engle, R. A., Smith, M. S., & Hughes, E. K. (2008). Orchestrating productive mathematical discussions: Five practices for helping teachers move beyond show and tell. *Mathematical Thinking and Learning, 10*(4), 313–340.

Stein, M. K., Grover, B., & Henningsen, M. (1996). Building student capacity for mathematical thinking and reasoning: An analysis of mathematical tasks used in reform classrooms. *American Educational Research Journal, 33*(2), 455–488.

Stein, M. K., & Smith, M. S. (2011). *5 practices for orchestrating productive mathematics discussions.* Reston, VA: NCTM.

Stenmark, J. (Ed.) (1991). *Mathematics assessment: Myths, models, good questions, and practical suggestions.* Reston, VA: NCTM.

Swan, M. (2006). *Collaborative learning in mathematics: A challenge to our beliefs and practices.* London: National Institute for Advanced and Continuing Education (NIACE) for the National Research and Development Centre for Adult Literacy and Numeracy (NRDC). ISBN 1-86201-311-X.

Swan, M., & Burkhardt, H. (2014) Lesson design for formative assessment. *Educational Designer,* 2(7), Downloaded from http://www.educationaldesigner.org/ed/volume2/issue7/article24/index.htm.

Teaching for Robust Understanding (TRU) Framework web site, https://truframework.org.

Van de Walle, J., Karp, K., & Bay-Williams, J. (2018). *Elementary and middle school mathematics: Teaching developmentally.* New York: Pearson.

Wertheimer, M. (1945). *Productive thinking.* New York: Harper & Row.

Wiliam, D. (2017). Assessment for learning: Meeting the challenge of Implementation. *Assessment in Education: Principles, Policy & Practice.* Downloaded December 30, 2017, from https://doi.org/10.1080/0969594X.2017.1401526.

Wiliam, D., & Thompson, M. (2007). Integrating assessment with instruction: What will it take to make it work? In C. A. Dwyer (Ed.), *The future of assessment: Shaping teaching and learning* (pp. 53–82). Mahwah, NJ: Lawrence Erlbaum.

Zwiers, J., Dieckmann, J., Rutherford-Quach, S., Daro, V., Skarin, R., Weiss, S., & Malamut, J. (2017). *Principles for the design of mathematics curricula: Promoting language and content development.* Retrieved from Stanford University, UL/SCALE website: http://ell.stanford.edu/content/mathematics-resources-additional-resources.